Science Notebook

Glencoe Science

Life Science

Consultant
Douglas Fisher, Ph.D.

New York, New York   Columbus, Ohio   Chicago, Illinois   Peoria, Illinois   Woodland Hills, California

About the Consultant

Douglas Fisher, Ph.D., is a Professor in the Department of Teacher Education at San Diego State University. He is the recipient of an International Reading Association Celebrate Literacy Award as well as a Christa McAuliffe award for Excellence in Teacher Education. He has published numerous articles on reading and literacy, differentiated instruction, and curriculum design as well as books, such as *Improving Adolescent Literacy: Strategies at Work* and *Responsive Curriculum Design in Secondary Schools: Meeting the Diverse Needs of Students*. He has taught a variety of courses in SDSU's teacher-credentialing program as well as graduate-level courses on English language development and literacy. He also has taught classes in English, writing, and literacy development to secondary school students.

The McGraw·Hill Companies

Copyright © by the McGraw-Hill Companies, Inc. All rights reserved. Permission is granted to reproduce the material contained herein on the condition that such material be reproduced only for classroom use; be provided to students, teachers, and families without charge; and be used solely in conjunction with *Life Science*. Any other reproduction, for use or sale, is prohibited without prior written permission of the publisher.

Send all inquiries to:
Glencoe/McGraw-Hill
8787 Orion Place
Columbus, Ohio 43240-4027

ISBN 0-07-874567-5

Printed in the United States of America

2 3 4 5 6 7 8 9   024   08 07 06

Table of Contents

Note-Taking Tips v

Using Your Science Notebook vi

Chapter 1 Exploring and Classifying Life
Chapter Preview 1
1-1 2
1-2 5
1-3 8
1-4 11
Wrap-Up 14

Chapter 2 Cells
Chapter Preview 15
2-1 16
2-2 19
2-3 22
Wrap-Up 26

Chapter 3 Cell Processes
Chapter Preview 27
3-1 28
3-2 31
3-3 34
Wrap-Up 38

Chapter 4 Cell Reproduction
Chapter Preview 39
4-1 40
4-2 43
4-3 46
Wrap-Up 50

Chapter 5 Heredity
Chapter Preview 51
5-1 52
5-2 55
5-3 58
Wrap-Up 62

Chapter 6 Adaptations over Time
Chapter Preview 63
6-1 64
6-2 67
6-3 70
Wrap-Up 74

Chapter 7 Bacteria
Chapter Preview 75
7-1 76
7-2 79
Wrap-Up 82

Chapter 8 Protists and Fungi
Chapter Preview 83
8-1 84
8-2 87
Wrap-Up 90

Chapter 9 Plants
Chapter Preview 91
9-1 92
9-2 95
9-3 98
Wrap-Up 102

Chapter 10 Plant Reproduction
Chapter Preview 103
10-1 104
10-2 107
10-3 110
Wrap-Up 114

Chapter 11 Plant Processes
Chapter Preview 115
11-1 116
11-2 118
Wrap-Up 122

Chapter 12 Introduction to Animals
Chapter Preview 123
12-1 124
12-2 127
12-3 130
Wrap-Up 134

Chapter 13 Mollusks, Worms, Arthropods, Echinoderms
Chapter Preview 135
13-1 136
13-2 139
13-3 142
13-4 145
Wrap-Up 148

Life Science

Table of Contents

Chapter 14 Fish, Amphibians, and Reptiles
Chapter Preview 149
14-1 150
14-2 153
14-3 156
14-4 159
Wrap-Up 162

Chapter 15 Birds and Mammals
Chapter Preview 163
15-1 164
15-2 167
Wrap-Up 170

Chapter 16 Animal Behavior
Chapter Preview 171
16-1 172
16-2 175
Wrap-Up 178

Chapter 17 Structure and Movement
Chapter Preview 179
17-1 180
17-2 183
17-3 186
Wrap-Up 190

Chapter 18 Nutrients and Digestion
Chapter Preview 191
18-1 192
18-2 195
Wrap-Up 198

Chapter 19 Circulation
Chapter Preview 199
19-1 200
19-2 203
19-3 206
Wrap-Up 210

Chapter 20 Respiration and Excretion
Chapter Preview 211
20-1 212
20-2 215
Wrap-Up 218

Chapter 21 Control and Coordination
Chapter Preview 219
21-1 220
21-2 223
Wrap-Up 226

Chapter 22 Regulation and Reproduction
Chapter Preview 227
22-1 228
22-2 231
22-3 234
Wrap-Up 238

Chapter 23 Immunity and Disease
Chapter Preview 239
23-1 240
23-2 243
23-3 246
Wrap-Up 250

Chapter 24 Interactions of Life
Chapter Preview 251
24-1 252
24-2 255
24-3 258
Wrap-Up 262

Chapter 25 The Nonliving Environment
Chapter Preview 263
25-1 264
25-2 267
25-3 270
Wrap-Up 274

Chapter 26 Ecosystems
Chapter Preview 275
26-1 276
26-2 279
23-3 282
23-4 285
Wrap-Up 286

Chapter 27 Conserving Resources
Chapter Preview 287
27-1 288
27-2 291
27-3 294
Wrap-Up 298

Academic Vocabulary 299

Note-Taking Tips

Your notes are a reminder of what you learned in class. Taking good notes can help you succeed in science. These tips will help you take better notes.

- Be an active listener. Listen for important concepts. Pay attention to words, examples, and/or diagrams your teacher emphasizes.
- Write your notes as clearly and concisely as possible. The following symbols and abbreviations may be helpful in your note-taking.

Word or Phrase	Symbol or Abbreviation	Word or Phrase	Symbol or Abbreviation
for example	e.g.	and	+
such as	i.e.	approximately	≈
with	w/	therefore	∴
without	w/o	versus	vs

- Use a symbol such as a star (★) or an asterisk (*) to emphasis important concepts. Place a question mark (?) next to anything that you do not understand.
- Ask questions and participate in class discussion.
- Draw and label pictures or diagrams to help clarify a concept.

Note-Taking Don'ts

- **Don't** write every word. Concentrate on the main ideas and concepts.
- **Don't** use someone else's notes—they may not make sense.
- **Don't** doodle. It distracts you from listening actively.
- **Don't** lose focus or you will become lost in your note-taking.

Using Your Science Notebook

This note-taking guide is designed to help you succeed in learning science content. Each chapter includes:

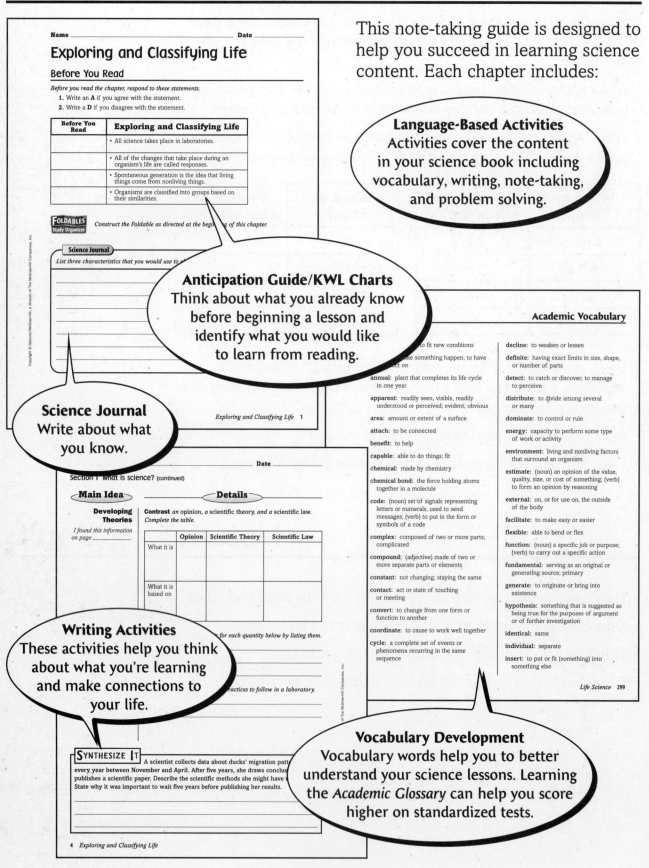

Language-Based Activities
Activities cover the content in your science book including vocabulary, writing, note-taking, and problem solving.

Anticipation Guide/KWL Charts
Think about what you already know before beginning a lesson and identify what you would like to learn from reading.

Science Journal
Write about what you know.

Writing Activities
These activities help you think about what you're learning and make connections to your life.

Vocabulary Development
Vocabulary words help you to better understand your science lessons. Learning the *Academic Glossary* can help you score higher on standardized tests.

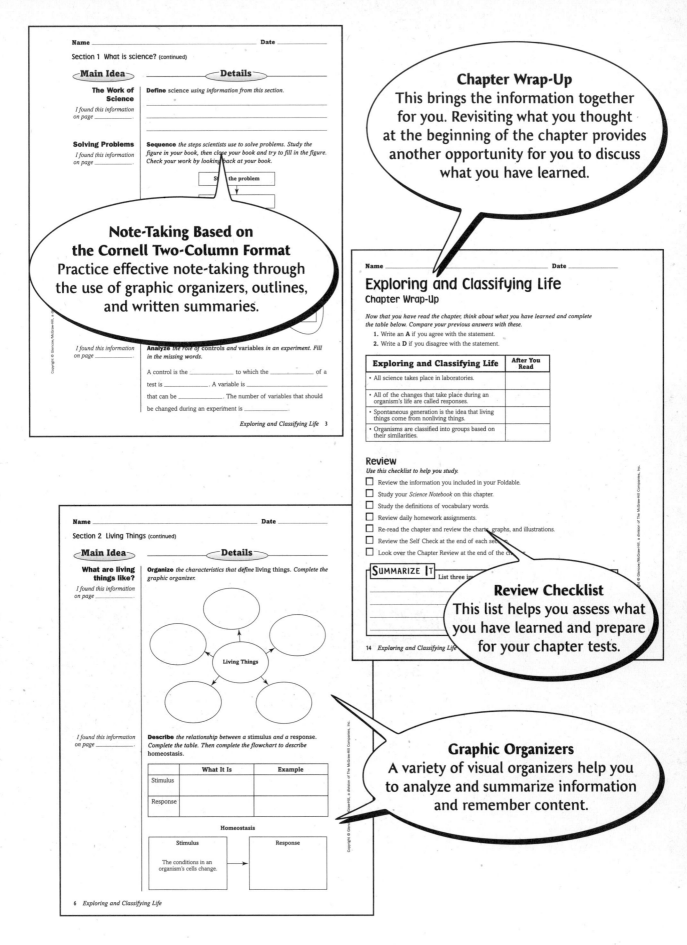

Exploring and Classifying Life

Before You Read

Before you read the chapter, respond to these statements.

1. Write an **A** if you agree with the statement.
2. Write a **D** if you disagree with the statement.

Before You Read	Exploring and Classifying Life
	• All science takes place in laboratories.
	• All of the changes that take place during an organism's life are called responses.
	• Spontaneous generation is the idea that living things come from nonliving things.
	• Organisms are classified into groups based on their similarities.

 Construct the Foldable as directed at the beginning of this chapter.

Science Journal

List three characteristics that you would use to classify underwater life.

_____
_____
_____
_____
_____
_____

Exploring and Classifying Life 1

Name _____   Date _____

Exploring and Classifying Life
Section 1 What is science?

Scan *the list below to preview Section 1 of your book.*
- Read all section headings.
- Read all bold words.
- Read all charts and graphs.
- Think about what you already know about how to solve problems.

Write *three facts you discovered about* scientific methods *as you scanned the section.*

1. _____
2. _____
3. _____

Review Vocabulary

experiment

Write *a paragraph describing scientific methods. Use all of the vocabulary words in your description. Underline each vocabulary word.*

New Vocabulary

scientific methods
hypothesis
control
variable
theory
law

Academic Vocabulary

reject

_____
_____
_____
_____
_____
_____
_____
_____
_____
_____
_____
_____

2 Exploring and Classifying Life

Name _____ Date _____

Section 1 What is science? (continued)

Main Idea	Details
The Work of Science *I found this information on page* _____.	**Define** science *using information from this section.* _____ _____ _____ _____
Solving Problems *I found this information on page* _____.	**Sequence** *the steps scientists use to solve problems. Study the figure in your book, then close your book and try to fill in the figure. Check your work by looking back at your book.* 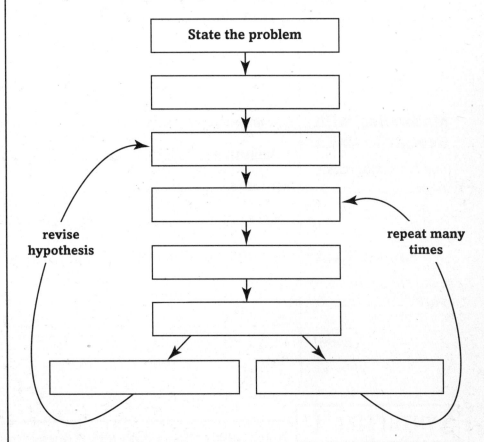
I found this information on page _____.	**Analyze** *the role of* controls *and* variables *in an experiment. Fill in the missing words.* A control is the _____ to which the _____ of a test is _____. A variable is _____ that can be _____. The number of variables that should be changed during an experiment is _____.

Exploring and Classifying Life 3

Name _____   Date _____

Section 1 What is science? (continued)

Main Idea — Details

Developing Theories

I found this information on page _____.

Contrast *an opinion, a scientific theory, and a scientific law. Complete the table.*

	Opinion	Scientific Theory	Scientific Law
What it is			
What it is based on			

Measuring with Scientific Units

I found this information on page _____.

Summarize *the metric units for each quantity below by listing them.*

Length: _____

Volume: _____

Mass: _____

Safety First

I found this information on page _____.

Identify *two important safety practices to follow in a laboratory.*

1. _____

2. _____

SYNTHESIZE IT

A scientist collects data about ducks' migration patterns every year between November and April. After five years, she draws conclusions and publishes a scientific paper. Describe the scientific methods she might have used. State why it was important to wait five years before publishing her results.

_____
_____
_____

4 *Exploring and Classifying Life*

Name _____ Date _____

Exploring and Classifying Life
Section 2 Living Things

Predict what you will learn in Section 2. Read the title and main headings. List three topics that you predict will be discussed in the section.

1. _____
2. _____
3. _____

Review Vocabulary *Use* raw materials *in a sentence to show its scientific meaning.*

raw materials _____
_____

New Vocabulary *Find a sentence in Section 2 that uses each vocabulary term.*

organism _____
_____
_____

cell _____
_____
_____

homeostasis _____
_____
_____

Academic Vocabulary *Use a dictionary to define* chemical.

chemical _____
_____
_____

Exploring and Classifying Life

Name _____ Date _____

Section 2 Living Things (continued)

Main Idea

What are living things like?

I found this information on page _____.

Details

Organize the characteristics that define living things. Complete the graphic organizer.

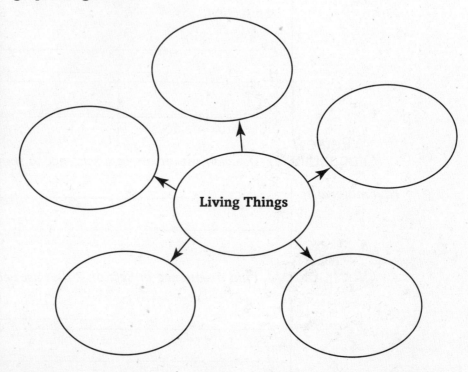

I found this information on page _____.

Describe the relationship between a *stimulus* and a *response*. Complete the table. Then complete the flowchart to describe homeostasis.

	What It Is	Example
Stimulus		
Response		

Homeostasis

Stimulus	Response
The conditions in an organism's cells change.	

6 *Exploring and Classifying Life*

Name _____ Date _____

Section 2 Living Things (continued)

Main Idea	**Details**

I found this information on page _____.

Contrast the ways organisms obtain energy in the table.

Organism	How It Obtains Energy
Plants	
Animals	
Bacteria in places sunlight cannot reach	

What do living things need?

I found this information on page _____.

Classify the needs of all living things. Complete the concept map.

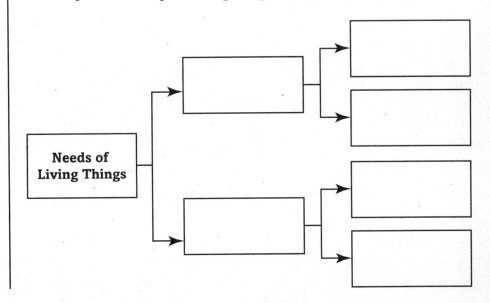

SUMMARIZE IT Choose one living thing and one nonliving thing with which you are familiar. Use the five characteristics of living things to explain how you know that each is living or nonliving. Complete the chart to organize your information.

Object	Has cells?	Uses energy?	Grows and develops?	Responds to stimuli?	Reproduces?

Exploring and Classifying Life 7

Exploring and Classifying Life

Section 3 Where does life come from?

Skim Section 3, and write three questions that you have.

1. _____
2. _____
3. _____

Review Vocabulary **Define** contaminate *and use it in an original sentence.*

contaminate _____

_____

_____

New Vocabulary *Write the vocabulary term that matches each definition.*

_____ | the idea that living things come from nonliving things

_____ | the idea that living things come only from other living things

Academic Vocabulary *Use a dictionary to define* estimate *as both a noun and a verb.*

estimate

noun: _____

_____

verb: _____

_____

Name _____ Date _____

Section 3 Where does life come from? (continued)

Main Idea | Details

Life Comes from Life

I found this information on page _____ .

Contrast the theories of spontaneous generation *and* biogenesis. Complete the table.

	Spontaneous Generation	Biogenesis
Source of life		

I found this information on page _____ .

Sequence experiments that were conducted about the theory of spontaneous generation. Complete the time line.

1800s
Who: _____
What: _____
_____

1700s
Who: _____
What: _____
_____

1668
Who: _____
What: _____
_____

Life's Origins

I found this information on page _____ .

Complete key events in the evolution of life on Earth. Identify the event that scientists hypothesize occurred at each time.

about 5 billion years ago: _____
_____

about 4.6 billion years ago: _____

more than 3.5 billion years ago: _____

Exploring and Classifying Life   9

Name _____ Date _____

Section 3 Where does life come from? (continued)

Main Idea	**Details**

Life's Origins

I found this information on page _____.

Organize information about Oparin's hypothesis. **Complete the outline.**

I. Oparin's hypothesis of Earth's early atmosphere composition

    A. _____

    B. _____

    C. _____

    D. _____

II. What happened in the atmosphere

    A. _____

    B. _____

Complete the graphic organizer summarizing Stanley Miller and Harold Urey's experiment.

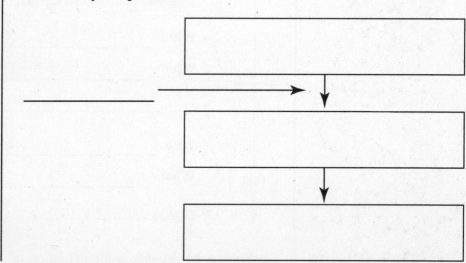

CONNECT IT Scientists' theories of the origin of life have changed over time. How do these changes show the use of scientific methods?

_____
_____
_____
_____

10 *Exploring and Classifying Life*

Exploring and Classifying Life

Section 4 How are living things classified?

Read the What You'll Learn *statements for Section 4. Rewrite each statement as a question. As you read, look for the responses to your questions.*

1. _____
2. _____
3. _____
4. _____

Review Vocabulary **Describe** *how an organism's common name is different from its scientific name.*

common name _____
_____
_____

New Vocabulary *Read the definitions below. Write the vocabulary term that matches each definition.*

_____ first and largest category used to classify organisms

_____ evolutionary history of an organism

_____ group of similar species

_____ two-word scientific naming system

Academic Vocabulary **Define** *similar using a dictionary.*

similar _____
_____
_____

Exploring and Classifying Life 11

Name _____ Date _____

Section 4 How are living things classified? (continued)

Main Idea	**Details**

Classification

I found this information on page _____.

Contrast *historic classification systems. Identify the categories or criteria used in each system.*

	Early classification	Aristotle	Linnaeus
Categories or criteria			

I found this information on page _____.

Summarize *the 6 types of information that modern scientists use to determine an organism's* phylogeny.

1. _____
2. _____
3. _____
4. _____
5. _____
6. _____

I found this information on page _____.

Label *the groups used to classify organisms from least specific to most specific. Use the word bank to complete the diagram.*

class   genus   order   species
family   kingdom   phylum

12 *Exploring and Classifying Life*

Name _____ Date _____

Section 4 How are living things classified? (continued)

Main Idea

Details

Scientific Names

I found this information on page _____.

Summarize binomial nomenclature. *Complete the sentences.*

The first word of an organism's scientific name is its _____.

The second word might _____.

Identify four reasons the system of binomial nomenclature is useful.

1. _____
   _____

2. _____
   _____

3. _____
   _____

4. _____
   _____

Tools for Identifying Organisms

I found this information on page _____.

Distinguish between *a* field guide *and a* dichotomous key. *Complete the Venn diagram.*

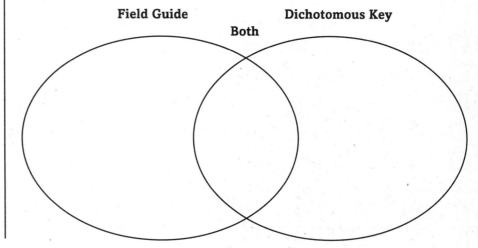

SYNTHESIZE IT Choose five similar plants or animals. Use what you know about their structures and features to develop your own dichotomous key to classify your choices. Use a dictionary to find the scientific name of each plant or animal to include in your key.

Exploring and Classifying Life 13

Name _____ Date _____

Exploring and Classifying Life
Chapter Wrap-Up

Now that you have read the chapter, think about what you have learned and complete the table below. Compare your previous answers with these.

1. Write an **A** if you agree with the statement.
2. Write a **D** if you disagree with the statement.

Exploring and Classifying Life	**After You Read**
• All science takes place in laboratories.	
• All of the changes that take place during an organism's life are called responses.	
• Spontaneous generation is the idea that living things come from nonliving things.	
• Organisms are classified into groups based on their similarities.	

Review

Use this checklist to help you study.

☐ Review the information you included in your Foldable.
☐ Study your *Science Notebook* on this chapter.
☐ Study the definitions of vocabulary words.
☐ Review daily homework assignments.
☐ Re-read the chapter and review the charts, graphs, and illustrations.
☐ Review the Self Check at the end of each section.
☐ Look over the Chapter Review at the end of the chapter.

SUMMARIZE IT List three important ideas you learned in Chapter 1.

_____
_____
_____

Name _____ Date _____

Cells

Before You Read

Preview the chapter title, the section titles, and the section headings. List at least one idea for each section in each column.

K What I know	W What I want to find out

 Construct the Foldable as directed at the beginning of this chapter.

Science Journal

Write three questions that you would ask a scientist researching cancer cells.

_____
_____
_____
_____
_____
_____
_____

Cells 15

Name _____ Date _____

Cells
Section 1 Cell Structure

Skim Section 1. Write two questions that come to mind.

1. _____
2. _____

Review Vocabulary — **Write** sentences using the Review Vocabulary and New Vocabulary words. Use two or more of the vocabulary words in each sentence.

photosynthesis _____

New Vocabulary

cell membrane _____

cytoplasm _____

cell wall _____

organelle _____

nucleus _____

chloroplast _____

mitochondrion _____

ribosome _____

endoplasmic reticulum _____

Golgi body _____

tissue _____

organ _____

Academic Vocabulary — **Write** sentences using function as a noun and as a verb.

function

Noun: _____

Verb: _____

Name _____ Date _____

Section 1 Cell Structure (continued)

Main Idea | **Details**

Common Cell Traits

I found this information on page _____.

Define cell *by completing the following statement.*

A cell is _____
_____.

I found this information on page _____.

Model *a* prokaryotic cell *and a* eukaryotic cell. *Show the difference between the two types.*

Prokaryotic Cell	Eukaryotic Cell

Cell Organization

I found this information on page _____.

Organize *information about eukaryotic cell parts in the table.*

Part	Description
Cell wall	
Nucleus	
Chloroplast	
Mitochondria	
Ribosomes	
Endoplasmic reticulum	
Golgi bodies	
Lysosomes	

Cells   17

Name _____ Date _____

Section 1 Cell Structure (continued)

Main Idea	**Details**

From Cell to Organism

I found this information on page _____.

Sequence *the following terms from simplest (at the top) to most complex in the chart below. Define each term and provide an example.*

tissue   organism   cell   organ system   organ

Term: _____   Example: _____
Definition: _____

↓

Term: _____   Example: _____
Definition: _____

↓

Term: _____   Example: _____
Definition: _____

↓

Term: _____   Example: _____
Definition: _____

↓

Term: _____   Example: _____
Definition: _____

SYNTHESIZE IT Compare and contrast animal and plant cells.

_____
_____
_____

18 Cells

Name _____ Date _____

Cells
Section 2 Viewing Cells

Predict *three things that might be discussed in this section after reading its headings.*

1. _____

2. _____

3. _____

Review Vocabulary **Use** *magnify in a sentence.*

magnify
_____
_____

New Vocabulary *Find a sentence in Section 2 in which* **cell theory** *is used and write it here.*

cell theory
_____
_____

Academic Vocabulary **Define** *compound as an adjective. Use a dictionary if you need to.*

compound
_____
_____

Locate and write a sentence in Section 2 in which the word **compound** *is used as an adjective.*

_____
_____
_____

Cells 19

Name _____ Date _____

Section 2 Viewing Cells (continued)

Main Idea | Details

Magnifying Cells

I found this information on page _____.

Summarize information in your book to describe van Leeuwenhoek's microscope.

_____
_____
_____

I found this information on page _____.

Evaluate the total magnification of a microscope with a 10X eyepiece lens and a 43X objective lens. Write the equation for finding total magnification. Then use it to show your calculation.

total magnification =

_____

total magnification = _____

I found this information on page _____.

Compare compound microscopes with electron microscopes by completing the Venn diagram with at least seven facts.

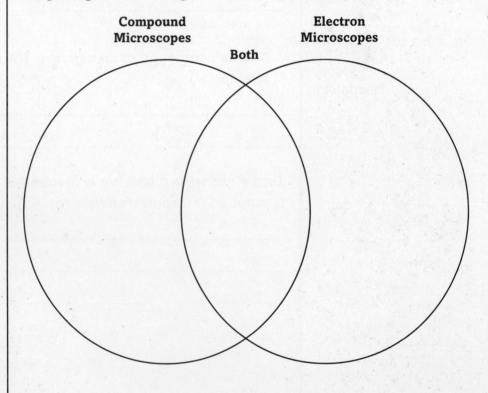

20 Cells

Name _____ Date _____

Section 2 Viewing Cells (continued)

Main Idea	**Details**

Cell Theory

I found this information on page _____.

Summarize discoveries made by scientists that led to the cell theory.

Robert Hooke _____
_____
_____

Matthias Schleiden _____
_____
_____

Theodor Schwann _____
_____
_____

Rudolf Virchow _____
_____
_____

I found this information on page _____.

List the 3 main principles of the cell theory.

1. _____
2. _____
3. _____

CONNECT IT Describe how the development of the cell theory shows that scientific beliefs can change over time. Use specific examples.

_____
_____
_____
_____
_____

Cells
Section 3 Viruses

Scan *Section 3 of this chapter. Write three questions based on headings in the section. Answer the questions as you read.*

1. _____
2. _____
3. _____

Review Vocabulary **Define** *disease using your book or a dictionary.*

disease _____
_____

New Vocabulary *Use your book to define each new vocabulary term.*

virus _____
_____
_____

host cell _____
_____

Academic Vocabulary *Use a dictionary to define* **apparent.**

apparent _____
_____

Explain what the following sentence means.

The virus is still in your body's cells, but it is hiding and doing no *apparent* harm.

_____
_____
_____

Name _____ Date _____

Section 3 Viruses (continued)

Main Idea | Details

What are viruses?

I found this information on page _____.

Organize information about viruses by completing the outline.

Viruses
  I. Definition: _____
  _____
  II. Description:
     A. Size: _____
     B. Shapes: _____
  III. Diseases caused by viruses
     A. _____ C. _____
     B. _____ D. _____

How do viruses multiply?

I found this information on page _____.

Summarize what a virus needs to reproduce.
_____

Distinguish between an active virus and a latent virus.

A(n) _____ enters a host cell, immediately causes the cell to make new viruses, and destroys the cell.

A(n) _____ enters a host cell, but does not immediately make new viruses or destroy the cell.

Sequence the events when an active virus enters a host cell.

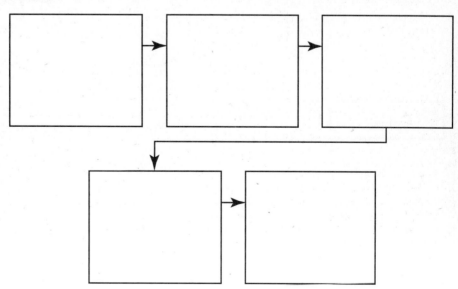

Cells 23

Name _____ Date _____

Section 3 Viruses (continued)

Main Idea | Details

How do viruses affect organisms?
I found this information on page _____.

Define bacteriophage *and explain what it does to a* bacterium.
_____
_____
_____
_____
_____

Fighting Viruses
I found this information on page _____.

Sequence *the steps by which* interferons *work*.

```
┌─────────────────────────────────────┐
│                                     │
└─────────────────────────────────────┘
                  ↓
┌─────────────────────────────────────┐
│                                     │
└─────────────────────────────────────┘
                  ↓
┌─────────────────────────────────────┐
│                                     │
└─────────────────────────────────────┘
```

Research with Viruses
I found this information on page _____.

Summarize *how scientists use viruses in* gene therapy.
_____
_____
_____
_____

CONNECT IT Describe why it is not a good idea to take antibiotics for a cold.
_____
_____
_____
_____

Tie It Together

A scientist is researching an unknown disease. After examining the disease-causing agent with a compound microscope and testing it in various ways, she has decided that the disease should be treated with an antibiotic drug to disrupt its membrane and prevent it from reproducing. Describe what is causing the disease and how you know.

Name _____ Date _____

Cells Chapter Wrap-Up

Review the ideas you listed at the beginning of the chapter. Cross out any incorrect information in the first column. Then complete the table by filling in the third column.

K What I know	W What I want to find out	L What I learned

Review
Use this checklist to help you study.

- ☐ Review the information you included in your Foldable.
- ☐ Study your *Science Notebook* on this chapter.
- ☐ Study the definitions of vocabulary words.
- ☐ Review daily homework assignments.
- ☐ Re-read the chapter and review the charts, graphs, and illustrations.
- ☐ Review the Self Check at the end of each section.
- ☐ Look over the Chapter Review at the end of the chapter.

SUMMARIZE IT What are the three most important ideas in the chapter?

_____
_____
_____
_____
_____

Name _____ Date _____

Cell Processes

Before You Read

Before you read the chapter, respond to these statements.

1. Write an **A** if you agree with the statement.
2. Write a **D** if you disagree with the statement.

Before You Read	Cell Processes
	• Matter is made up of atoms.
	• All substances chemically combine when they are mixed together.
	• Energy is always needed to move material across a cell membrane.
	• Plants can convert light energy into chemical energy.

 Construct the Foldable as directed at the beginning of this chapter.

Science Journal

Describe two ways in which you think plants get food and energy.

_____
_____
_____
_____
_____
_____

Cell Processes
Section 1 Chemistry of Life

Predict *what you will learn in Section 1 after reading the headings and looking at the diagrams.*

1. _____
2. _____
3. _____

Review Vocabulary **Define** *cell to show its scientific meaning.*

cell _____
_____

New Vocabulary *Find each term in Section 1 and write the sentence where it is used.*

mixture _____
_____
_____

organic compound _____
_____
_____

enzyme _____
_____
_____

inorganic compound _____
_____
_____

Academic Vocabulary *Use a dictionary to define chemical bond.*

chemical bond _____
_____

Name _____ Date _____

Section 1 Chemistry of Life (continued)

Main Idea | Details

The Nature of Matter

I found this information on page _____.

Compare elements *and* compounds *by completing the chart below.*

	Elements	Compounds
Number of types of atom		
Example		

I found this information on page _____.

Classify *each characteristic of compounds as* ionic, molecular, *or* both.

_____ has positively and negatively charged ions

_____ share outermost electrons to bond

_____ salt

_____ sugar

_____ involved in many life processes

_____ have different properties than the elements from which they are made

Mixtures

I found this information on page _____.

Compare mixtures, solutions, *and* suspensions. *Complete the statements below.*

A mixture is _____

_____

Both solutions and suspensions _____

_____

In a solution, _____

_____

In a suspension, _____

_____

Cell Processes 29

Name _____   Date _____

Section 1 Chemistry of Life (continued)

Main Idea / Details

Organic Compounds

I found this information on page _____.

Summarize the functions of the 4 main organic compounds.

Organic Compounds in Living Things	
Compound	Function
Carbohydrates	
Lipids	
Proteins	
Nucleic acids	

Inorganic Compounds

I found this information on page _____.

Compare and contrast characteristics of organic and inorganic compounds by completing the table below.

Characteristic	Organic	Inorganic
Contains carbon?		
Role in living things		

I found this information on page _____.

Identify three ways that water is important to living things.

1. _____

2. _____

3. _____

30 Cell Processes

Name _____ Date _____

Cell Processes
Section 2 Moving Cellular Materials

Skim *Section 2. List three headings you would use to make an outline of this section.*

1. _____

2. _____

3. _____

Review Vocabulary *Define* cytoplasm *to show its scientific meaning.*

cytoplasm
_____
_____

New Vocabulary *Write the vocabulary term that matches each definition.*

_____ movement of substances through a cell membrane without the use of energy

_____ occurs when molecules of one substance are spread evenly throughout another substance

_____ energy-requiring process in which transport proteins bind with particles and move them through a cell membrane

_____ process by which a cell takes in a substance by surrounding it with the cell membrane

_____ process by which vesicles release their contents outside the cell

_____ type of passive transport in which molecules move from where there are more of them to where there are fewer of them

_____ type of passive transport that occurs when water diffuses through a cell membrane

Academic Vocabulary *Use a dictionary to define the term* facilitate.

facilitate
_____
_____

Name _____ Date _____

Section 2 Moving Cellular Materials (continued)

Main Idea | Details

I found this information on page _____.

Create *a diagram that shows how oxygen diffuses from air sacs in the lungs to red blood cells.*

I found this information on page _____.

Write a short caption on how oxygen moves from the lungs to toe cells.

_____
_____
_____

Complete *the concept map of osmosis.*

- is a type of _____.
- does not require _____.
- involves the movement of _____ through the cell membrane.
- occurs in both plant and animal _____.

Osmosis

I found this information on page _____.

List *three facts about facilitated diffusion.*

1. _____
2. _____
3. _____

32   Cell Processes

Name _____ Date _____

Section 2 Moving Cellular Materials (continued)

Main Idea	**Details**

Active Transport

I found this information on page _____.

Sequence *the process of how* active transport *moves materials into the cell.*

1. _____

2. _____

3. _____

I found this information on page _____.

Compare and contrast *facilitated diffusion and active transport by writing yes or no in each box of the chart.*

	Facilitated Diffusion	Active Transport
Uses transport proteins?		
Transports materials across cell membrane?		
Requires energy?		
Able to move materials from an area with less of the material to an area with more of the material?		

Endocytosis and Exocytosis

I found this information on page _____.

Complete *the table to identify the processes involved in moving very large particles in and out of cells.*

	Process	Description
Materials entering cell		
Materials being expelled from cell		

Cell Processes
Section 3 Energy for Life

Scan *Section 3 of your book. Write three things you think you will learn about in this section.*

1. _____

2. _____

3. _____

Review Vocabulary **Define** mitochondrion *to show its scientific meaning.*

mitochondrion _____

New Vocabulary *Read the definitions below. Write the vocabulary term that matches the definition in the blank to the left.*

_____ process by which producers and consumers release stored energy from food molecules

_____ process by which oxygen-lacking cells and some one-celled organisms release small amounts of energy from glucose molecules and produce wastes such as alcohol, carbon dioxide, and lactic acid

_____ process by which plants and many other producers use light energy to produce a simple sugar from carbon dioxide and water and give off oxygen

_____ total of all chemical reactions in an organism

Academic Vocabulary *Use a dictionary to define* obtain.

obtain _____

34 *Cell Processes*

Name _____ Date _____

Section 3 Energy for Life (continued)

Main Idea | Details

Trapping and Using Energy

I found this information on page _____.

Model *a chemical reaction in which an* enzyme *changes two smaller molecules into one larger molecule.*

I found this information on page _____.

Complete *the table on the different materials and their roles in* photosynthesis.

Material	Role in Photosynthesis
Water	
Carbon dioxide	
	products of photosynthesis
Chlorophyll	

I found this information on page _____.

Analyze *why photosynthesis is important to animals.*

_____
_____
_____
_____
_____
_____

Cell Processes 35

Name _____ Date _____

Section 3 Energy for Life (continued)

Main Idea	Details
I found this information on page _____.	**Summarize** the process of respiration. *State what is broken down and what the products are.* _____ _____ _____
I found this information on page _____.	**Compare** fermentation *with respiration.*

Comparing Fermentation and Respiration		
Process	Fermentation	Respiration
What gets broken down?		
Where does breakdown occur?		
Is energy released?		
What wastes are produced?	if insufficient O_2 in muscle cells: _____ ; in yeast cells: _____	

SYNTHESIZE IT Describe the relationship between plants and animals. Use the listed terms in your description.

carbon dioxide   consumer   energy   oxygen   photosynthesis   producer   respiration

_____
_____
_____
_____

36   Cell Processes

Name _____ **Date** _____

Tie It Together

Suppose that you are small enough to be able to move around within the cytoplasm of a cell. Write a story about what it might be like to move through the cell membrane, including the method the cell would use to let you in. Explain why this is the best method.

_____
_____
_____
_____
_____
_____
_____
_____
_____
_____
_____
_____
_____
_____
_____
_____
_____
_____
_____
_____
_____
_____
_____
_____
_____

Name _____ Date _____

Cell Processes Chapter Wrap-Up

Now that you have read the chapter, think about what you have learned and complete the table below. Compare your previous answers with these.

1. Write an **A** if you agree with the statement.
2. Write a **D** if you disagree with the statement.

Cell Processes	**After You Read**
• Matter is made up of atoms.	
• All substances chemically combine when they are mixed together.	
• Energy is always needed to move material across a cell membrane.	
• Plants can convert light energy into chemical energy.	

Review

Use this checklist to help you study.

☐ Review the information you included in your Foldable.
☐ Study your *Science Notebook* on this chapter.
☐ Study the definitions of vocabulary words.
☐ Review daily homework assignments.
☐ Re-read the chapter and review the charts, graphs, and illustrations.
☐ Review the Self Check at the end of each section.
☐ Look over the Chapter Review at the end of the chapter.

SUMMARIZE IT List three important ideas in the chapter.

_____
_____
_____
_____

Cell Reproduction

Before You Read

Before you read the chapter, respond to these statements.

1. Write an **A** if you agree with the statement.
2. Write a **D** if you disagree with the statement.

Before You Read	Cell Reproduction
	• One-celled organisms reproduce through cell division.
	• Every living organism has a life cycle.
	• All organisms reproduce sexually.
	• Most of the cells formed in your body do not contain genetic material.

Construct the Foldable as directed at the beginning of this chapter.

Science Journal

Write three things that you know about how and why cells reproduce.

_____
_____
_____
_____
_____
_____

Cell Reproduction
Section 1 Cell Division and Mitosis

Skim Section 1 of your book. Read the headings, illustrations, and captions. Write three questions that come to mind as you skim the section.

1. _____
2. _____
3. _____

Review Vocabulary

Define nucleus *to show its scientific meaning.*

nucleus _____

New Vocabulary

Locate sentences in your book that use each of the following terms. Write each sentence here, and give the page on which you found it.

mitosis _____

chromosome _____

asexual reproduction _____

Academic Vocabulary

Use a dictionary to write a scientific definition of the term cycle. *Then find a sentence in this section that defines the* cell cycle, *and write it here.*

cycle _____

40 Cell Reproduction

Name _____ Date _____

Section 1 Cell Division and Mitosis (continued)

Main Idea | Details

Why is cell division important?

I found this information on page _____.

Identify the 3 reasons cell division is important.

1. _____
2. _____
3. _____

The Cell Cycle

I found this information on page _____.

Summarize information about interphase in eukaryotic cells in the following paragraph.

Interphase is the _____ part of the cell cycle. During interphase, cells _____ and _____. During interphase, cells that are still dividing copy their _____ and prepare for _____. Cells no longer dividing are _____.

Mitosis

I found this information on page _____.

Sequence the steps of mitosis, and write a short description of what takes place in each phase.

1. _____
   _____
2. _____
   _____
3. _____
   _____
4. _____
   _____
5. _____
   _____
6. _____
   _____

Cell Reproduction   41

Name _____ Date _____

Section 1 Cell Division and Mitosis (continued)

Main Idea | **Details**

I found this information on page _____.

Compare *mitosis in animals and plants. State if each feature exists in plant cells, animal cells, or both.*

Feature	Cell Type
Centrioles	
Spindle fibers	
Cell plate	
Cell wall	

I found this information on page _____.

Organize *important concepts about mitosis.*

1. Mitosis is the division of a _____.

2. Mitosis produces two new nuclei that are identical both to _____ and to _____.

3. A nucleus with 46 chromosomes that undergoes mitosis will produce _____ nuclei, each with _____ chromosomes.

Asexual Reproduction

I found this information on page _____.

Identify *the 3 forms of asexual reproduction described below.*

_____ the method by which bacteria reproduce

_____ new organism growing from body of the parent

_____ to regrow body parts that are lost or damaged

CONNECT IT A strawberry farmer wants to increase her crop without spending large amounts of money for new seeds. How can she take advantage of asexual reproduction to increase her crop?

_____

_____

_____

42 Cell Reproduction

Cell Reproduction
Section 2 Sexual Reproduction and Meiosis

Skim the headings and illustrations in Section 2. Write three things you think you will learn about in this section.

1. _____
2. _____
3. _____

Review Vocabulary **Define** organism *to show its scientific meaning.*

organism _____

New Vocabulary *Read the definitions below. Write the correct vocabulary term on the blank to the left.*

_____ in sexual reproduction, the joining of a sperm and egg

_____ new diploid cell formed when a sperm fertilizes an egg; will divide by mitosis and develop into a new organism

_____ sex cell formed in the female reproductive organs

_____ cell whose similar chromosomes occur in pairs

_____ reproductive process that produces haploid cells

_____ haploid sex cell formed in the male reproductive organs

_____ cells that have only half of each pair of chromosomes

_____ type of reproduction in which two sex cells join to form a zygote

Academic Vocabulary *Use a dictionary to define* process.

process _____

Cell Reproduction 43

Name _____ Date _____

Section 2 Sexual Reproduction and Meiosis (continued)

Main Idea — Details

Sexual Reproduction

I found this information on page _____.

Compare *characteristics of* human diploid *and* haploid cells *in the table below. Give examples of each type of cell.*

Types of Human Cells		
	Diploid	Haploid
Number of chromosomes		
Process that produces them		
Examples		

Meiosis and Sex Cells

I found this information on page _____.

Model *the 4 stages of* meiosis I *in the spaces below. Use the figure in your book to help you.*

Meiosis I	
Prophase I	Metaphase I
Anaphase I	Telophase I

44 Cell Reproduction

Name _____ Date _____

Section 2 Sexual Reproduction and Meiosis (continued)

Main Idea	**Details**

I found this information on page _____.

Model *what takes place inside a cell nucleus during* meiosis II *by drawing the 4 phases in the spaces below.*

Meiosis II	
Prophase II	Metaphase II
Anaphase II	Telophase II

I found this information on page _____.

Summarize *differences between* meiosis I *and* meiosis II *by writing a number, yes, or no in each box of the chart.*

	Meiosis I	Meiosis II
How many cells result?		
Is a haploid cell formed?		
Do chromatids separate?		

SYNTHESIZE IT

Fruit flies have eight chromosomes in their body cells. Mice have 40. How many chromosomes are there in each sex cell of these organisms?

_____

_____

Cell Reproduction 45

Cell Reproduction
Section 3 DNA

Scan *the list below to preview Section 3.*
- Read all section titles.
- Read all bold words.
- Look at all illustrations and their labels.
- Think about what you already know about DNA.

Review Vocabulary **Define** *heredity to show its scientific meaning.*

heredity _____

_____

New Vocabulary *Write the correct vocabulary term next to each definition.*

_____ deoxyribonucleic acid; a cell's heredity material; made up of two strands, each consisting of a sugar-phosphate backbone and nitrogen bases: adenine, thymine, guanine, and cytosine

_____ section of DNA that contains instructions for making specific proteins

_____ ribonucleic acid; type of nucleic acid that contains the sugar ribose, phosphates, and bases adenine, guanine, cytosine, and uracil

_____ any permanent change in a gene or chromosome of a cell; may be beneficial, harmful, or have little effect on an organism

Academic Vocabulary *The word* **code** *can be used as a noun or as a verb. Write a definition for its use as a noun and as a verb.*

code

Noun: _____

_____

Verb: _____

_____

46 *Cell Reproduction*

Name _____ Date _____

Section 3 DNA (continued)

Main Idea	Details
What is DNA?	**Identify** *the 4* nitrogen bases *found in* DNA.

I found this information on page _____.

1. _____   3. _____

2. _____   4. _____

I found this information on page _____.

Model *a section of a* DNA molecule, *showing its twisted-ladder structure. Label the* nitrogen bases, sugar, *and* phosphates. *Make sure the nitrogen bases in your drawing are correctly paired.*

[drawing box]

I found this information on page _____.

Summarize how DNA copies itself.

_____

_____

_____

_____

Genes

I found this information on page _____.

Complete *the following paragraph on the relationship of proteins and* genes.

Proteins are made up of long chains of _____.

Genes determine the _____ of _____

in a protein. Changing the _____ of the amino acids

makes a _____ protein.

Cell Reproduction 47

Name _____ Date _____

Section 3 DNA (continued)

Main Idea	**Details**

I found this information on page _____.

Complete the table on the 3 main kinds of RNA.

Type of RNA	Function
	carries the code to make proteins from the nucleus to the cytoplasm
transfer RNA (tRNA)	
	type of RNA contained in ribosomes

I found this information on page _____.

Complete the steps of protein production within a cell.

1. mRNA moves into the cytoplasm.
2. A(n) _____ attaches to it.
3. _____ molecules bring _____ to the ribosomes.
4. Nitrogen bases on the _____ temporarily _____ the nitrogen bases on the _____.
5. The same process occurs with another _____ molecule and the next portion of the _____ molecule.
6. The _____ attached to the two _____ molecules _____, beginning the formation of a protein.

Mutations

I found this information on page _____.

Describe how mutations can affect an organism.

_____
_____

CONNECT IT A man has a discolored area on the back of his hand. The doctor has assured him it is a harmless body cell mutation. Explain why the mutation probably will not appear in his children.

_____
_____

48 Cell Reproduction

Name _____ **Date** _____

Tie It Together

Draw an animal cell with six chromosomes.
Follow the chromosomes as they go through the steps of meiosis.
Show the chromosomes duplicating and separating, and describe the final end products.
Name each step in the process.
Show one way that a mutation might occur during the process.

Cell Reproduction

Name _____ Date _____

Cell Reproduction   Chapter Wrap-Up

Now that you have read the chapter, think about what you have learned and complete the table below. Compare your previous answers with these.

1. Write an **A** if you agree with the statement.
2. Write a **D** if you disagree with the statement.

Cell Reproduction	After You Read
• One-celled organisms reproduce through cell division.	
• Every living organism has a life cycle.	
• All organisms reproduce sexually.	
• Most of the cells formed in your body do not contain genetic material.	

Review

Use this checklist to help you study.

☐ Review the information you included in your Foldable.
☐ Study your *Science Notebook* on this chapter.
☐ Study the definitions of vocabulary words.
☐ Review daily homework assignments.
☐ Re-read the chapter and review the charts, graphs, and illustrations.
☐ Review the Self Check at the end of each section.
☐ Look over the Chapter Review at the end of the chapter.

SUMMARIZE IT   List three important ideas from this chapter.

_____
_____
_____
_____
_____

Heredity

Before You Read

Before you read the chapter, respond to these statements.

1. Write an **A** if you agree with the statement.
2. Write a **D** if you disagree with the statement.

Before You Read	Heredity
	• Offspring of an organism always have the same traits as the parents.
	• There may be more than two forms of a gene.
	• Some traits are determined by more than one gene.
	• Traits from one type of organism can be introduced into another type of organism.

 Construct the Foldable as directed at the beginning of the chapter.

Science Journal

Write three traits that you have and how you would determine how those traits were passed to you.

_____
_____
_____
_____
_____
_____

Name _____ Date _____

Heredity
Section 1 Genetics

Skim *Section 1 of the chapter. Write two questions that come to mind from reading the headings of this section.*

1. _____

2. _____

Review Vocabulary   **Define** meiosis.

meiosis   _____
_____
_____

New Vocabulary   *Write a paragraph describing* **heredity**. *Use the five vocabulary terms from the left in your paragraph.*

heredity   _____
genetics   _____
allele   _____
dominant   _____
recessive   _____

Write a paragraph describing **genotype**. *Use the five vocabulary terms from the left in your paragraph.*

Punnett square   _____
genotype   _____
phenotype   _____
homozygous   _____
heterozygous   _____
_____

Academic Vocabulary   *Use a dictionary to define* **physical**.

physical   _____

52   Heredity

Name _____ Date _____

Section 1 Genetics (continued)

Main Idea | Details

Inheriting Traits

I found this information on page _____.

Summarize what alleles *are and how they are inherited.*

_____
_____
_____
_____
_____

Mendel—The Father of Genetics

I found this information on page _____.

Identify three things Mendel did that made his work more useful than previous studies of *heredity.*

1. _____
_____

2. _____
_____

3. _____
_____

Genetics in a Garden

I found this information on page _____.

Analyze *one* trait *that Mendel studied.*

- Identify the *dominant* and *recessive* forms of the trait.
- Predict how an organism would look if it had two dominant alleles, two recessive alleles, or one of each allele.

Trait	
Dominant form	
Recessive form	
Two dominant alleles	
Two recessive alleles	
One of each allele	

Heredity 53

Name _____ Date _____

Section 1 Genetics (continued)

Main Idea

Genetics in a Garden

I found this information on page _____.

Details

Complete *the Punnett square for black and blond fur in a dog.*

	Black dog	
	B	b
Blond dog: b		
Blond dog: b		

Analyze *the Punnett square to complete the sentences.*

The black dog carries _____ black-fur traits. The blond dog carries _____ blond-fur traits. The chance that the offspring will have black fur is _____, or _____ in _____.

Summarize *Mendel's 3 principles of heredity.*

1. _____

2. _____

3. _____

I found this information on page _____.

CONNECT IT — A pea plant is *heterozygous* for purple flowers (Rr). A gardener crosses it with another pea plant with the same *genotype*. The recessive gene for this trait causes white flowers. Predict the possible genotypes and *phenotypes* for the offspring. Predict the percentage for each genotype and phenotype.

_____

_____

_____

54   Heredity

Name _____ Date _____

Heredity
Section 2 Genetics Since Mendel

Scan *the headings and illustrations in Section 2. Write two facts you learned about genetics as you scanned the section.*

1. _____

2. _____

Review Vocabulary **Define** *gene to show its scientific meaning.*

gene _____
_____

New Vocabulary *Define each vocabulary term.*

incomplete dominance _____
_____
_____

polygenic inheritance _____
_____
_____

sex-linked gene _____
_____
_____

Academic Vocabulary *Use a dictionary to define* intermediate. *Then rewrite the sentence below, using your definition.*

> When the allele for white four-o'clock flowers and the allele for red four-o'clock flowers combined, the result was an intermediate phenotype—pink flowers.

intermediate _____
_____
_____
_____

Heredity 55

Section 2 Genetics Since Mendel (continued)

Main Idea | Details

Incomplete Dominance

I found this information on page _____.

Draw *a Punnett square for red and white four-o'clock flowers showing the possible offspring. Use R for the allele for red flowers and R' for the allele for white flowers. In each section of the square, write the genotype and phenotype of the offspring.*

	Red four-o'clock	
	R	R
White four-o'clock R'		
R'		

Summarize incomplete dominance.

_____
_____

I found this information on page _____.

Analyze *how a gene with **multiple alleles** can produce more than three phenotypes. Use blood types as an example.*

_____
_____
_____
_____
_____

Polygenic Inheritance

I found this information on page _____.

Identify *how internal environment can affect the expression of a trait. Complete the flow chart.*

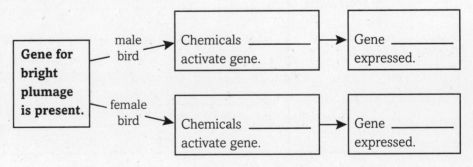

56 Heredity

Name _____ Date _____

Section 2 Genetics Since Mendel (continued)

Main Idea	Details
Human Genes and Mutations *I found this information on page* _____.	**Analyze** how chromosome disorders *occur*. A chromosome disorder occurs as a result of a _____ _____. It causes an organism to have _____ chromosomes than normal.
I found this information on page _____.	**Model** how two heterozygous parents who do not have a recessive disorder can have a child with the disorder. Use C for a dominant allele and c for a recessive allele. Mother's genotype: _____ Father's genotype: _____ → Child's genotype: _____
Sex-Linked Disorders *I found this information on page* _____.	**Complete** the statements about sex-linked traits. Sex-linked disorders usually result from _____ alleles on the _____ chromosome. A man will have the disorder when _____. A woman will have the disorder when _____ _____.
Pedigrees Trace Traits *I found this information on page* _____.	**Summarize** why pedigrees are useful to geneticists. _____ _____ _____

SYNTHESIZE IT Choose a trait described in Section 2, such as color-blindness, calico patterns in cats, or cystic fibrosis. Choose genotypes for two parents. Draw a pedigree starting with these parents. Continue your pedigree for two generations. Use Punnett squares to help you predict possible offspring.

Heredity 57

Heredity
Section 3 Biotechnology

Preview *the section title and headings. Write three questions that you would ask a modern geneticist after your preview.*

1. _____

2. _____

3. _____

Review Vocabulary *Use* DNA *in an original sentence to show its scientific meaning.*

DNA _____

New Vocabulary **Define** *genetic engineering.*

genetic engineering _____

Academic Vocabulary *Use a dictionary to define* insert *as a verb. Then find a sentence in Section 3 that uses the term or a form of the term.*

insert _____

Name _____   **Date** _____

Section 3 **Biotechnology** (continued)

Main Idea | Details

Genetic Engineering

I found this information on page _____ .

Distinguish *three uses for* genetic engineering.

1. _____
2. _____
3. _____

I found this information on page _____ .

Organize *information about* recombinant DNA. *Complete the graphic organizer.*

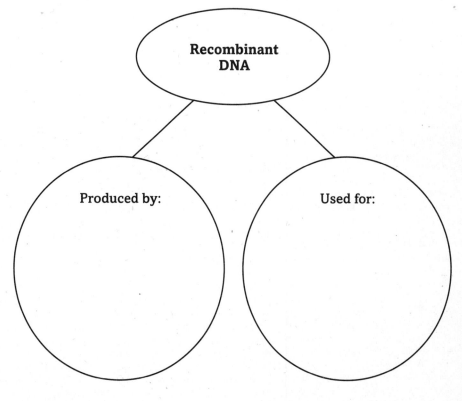

I found this information on page _____ .

Summarize *how* gene therapy *may be used in the future.*

_____
_____
_____
_____

Heredity 59

Name _____ **Date** _____

Section 3 Biotechnology (continued)

Main Idea | Details

Genetic Engineering

I found this information on page _____.

Create *a flow chart about gene therapy. Show how the gene gets into the body and what happens when it reaches the cells.*

[   ]

Summarize *each step of gene therapy in your model above.*

1. _____
2. _____
3. _____

I found this information on page _____.

Evaluate *the benefits and potential risks of genetic engineering of crop plants.*

Benefits	Risks

CONNECT IT Describe how viruses are useful tools in genetic engineering.

_____
_____
_____
_____

Name _____ **Date** _____

Tie It Together

Suppose that Gregor Mendel came to visit a modern genetics laboratory and you were asked to give him a tour. Write a report describing what you would show him and how you would explain modern genetics. Remember that he does not know the words gene *or* allele, *although he described "factors" that controlled traits.*

Name _____ Date _____

Heredity Chapter Wrap-Up

Now that you have read the chapter, think about what you have learned and complete the table below. Compare your previous answers with these.

1. Write an **A** if you agree with the statement.
2. Write a **D** if you disagree with the statement.

Heredity	**After You Read**
• Offspring of an organism always have the same traits as the parents.	
• There may be more than two forms of a gene.	
• Some traits are determined by more than one gene.	
• Traits from one type of organism can be introduced into another type of organism.	

Review

Use this checklist to help you study.

☐ Review the information you included in your Foldable.
☐ Study your *Science Notebook* on this chapter.
☐ Study the definitions of vocabulary words.
☐ Review daily homework assignments.
☐ Re-read the chapter and review the charts, graphs, and illustrations.
☐ Review the Self Check at the end of each section.
☐ Look over the Chapter Review at the end of the chapter.

SUMMARIZE IT Identify the three most important ideas in this chapter.

_____
_____
_____
_____

Name _____ Date _____

Adaptations over Time

Before You Read

Before you read the chapter, respond to these statements.

1. Write an **A** if you agree with the statement.
2. Write a **D** if you disagree with the statement.

Before You Read	Adaptations over Time
	• Traits acquired by an organism during its life can be passed on to its offspring.
	• Most evidence of evolution comes from fossils.
	• Organisms with traits best suited to their environment are more likely to survive and reproduce.
	• Humans share a common ancestor with other primates.

 Construct the Foldable as directed at the beginning of this chapter.

Science Journal

Pick a favorite plant or animal and list all the ways it is well-suited to its environment.

_____
_____
_____
_____
_____
_____
_____

Adaptations over Time 63

Name _____ Date _____

Adaptations over Time
Section 1 Ideas About Evolution

Predict three things that will be discussed in Section 1 as you scan the headings and illustrations.

1. _____
2. _____
3. _____

Review Vocabulary **Define** gene *using your book*.

gene

_____
_____
_____

New Vocabulary Write the correct term next to its definition.

_____ group of organisms that share similar characteristics and can reproduce among themselves, producing fertile offspring

_____ change in inherited characteristics over time

_____ process by which organisms with traits best suited to their environment are more likely to survive and reproduce

_____ inherited trait that makes an individual different from other members of its species

_____ any variation that makes an organism better suited to its environment

Academic Vocabulary *Use your book or a dictionary to define* hypothesis.

hypothesis

_____
_____
_____

64 *Adaptations over Time*

Name _____ Date _____

Section 1 Ideas About Evolution (continued)

Main Idea | Details

Early Models of Evolution

I found this information on page _____.

Identify why Lamarck's theory of evolution was not accepted.

_____
_____
_____

Darwin's Model of Evolution

I found this information on page _____.

Analyze Darwin's explanation of the origins of the 13 species of Galápagos finches. Fill in the missing words.

The Galápagos finches _____ for food. Those that had _____, _____ that allowed them to get food were able to _____ longer and _____ more. Over time, groups of finches became separate _____.

Natural Selection

I found this information on page _____.

State 5 main principles of natural selection.

1. _____

2. _____

3. _____

4. _____

5. _____

Variation and Adaptation

I found this information on page _____.

Compare and contrast variations and adaptations.

	Variation	Adaptation
Definition		
Examples		

Adaptations over Time   65

Name _____ Date _____

Section 1 Ideas About Evolution (continued)

Main Idea | Details

Variation and Adaptation

I found this information on page _____.

Complete the table explaining factors that can lead to changes in a population.

	What Happens	How It Leads to Change
Changes in Gene Sources		
Geographic Isolation		

The Speed of Evolution

I found this information on page _____.

Compare and contrast gradualism *and* punctuated equilibrium. *Select ideas from your reading to fill in the Venn diagram.*

Gradualism   Both   Punctuated Equilibrium

SYNTHESIZE IT Describe how natural selection can lead to the formation of a new species. Include factors such as migration and geographic isolation.

_____
_____
_____
_____
_____

Adaptations over Time

Adaptations over Time

Section 2 Clues About Evolution

Scan *Section 2 of your book. Then write two items in each of the boxes below.*

What I know about fossils	What I want to know about fossils

Review Vocabulary **Define** epoch *using your book.*

epoch _____

_____

New Vocabulary *Use your book to help you write the correct vocabulary term next to each definition.*

_____ a type of rock made from pieces of other rocks, minerals deposited from a solution, or plant and animal matter

_____ element that gives off a steady amount of radiation as it slowly changes to a nonradioactive element

_____ study of embryos and their development

_____ similar in structure, origin, or function

_____ structure that does not seem to have a function and that may once have functioned in the body of an ancestor

Academic Vocabulary *Use a dictionary to define* method.

method _____

_____

Name _____   Date _____

Section 2 Clues About Evolution (continued)

Main Idea	Details
Clues from Fossils *I found this information on page _____.*	**Create** *a concept map to summarize information about the* **Green River** *formation. Include information about* • where it is • what it was in the past • how fossils formed, and • what scientists learn from the fossils there.
Types of Fossils *I found this information on page _____.*	**Summarize** *the types of rock in which fossils are commonly found.* Most fossils are found in _____ rock. They are most often found in _____.

Determining a Fossil's Age
 	
I found this information on page _____.
	Organize *information about how scientists determine the age of fossils. Complete the outline.*
	   I. *Relative* dating
	      A. _____
	         _____
	      B. provides an estimate of a fossil's age by
	         _____
	   II. *Radiometric* dating
	      A. _____
	      B. Scientists estimate age by
	         _____
	         _____

Adaptations over Time

Name _____ Date _____

Section 2 Clues About Evolution (continued)

Main Idea	Details
Fossils and Evolution *I found this information on page* _____ .	**Create** a graphic organizer to identify what scientists learn from fossils. 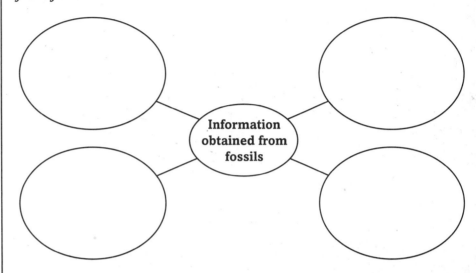
More Clues About Evolution *I found this information on page* _____ .	**Organize** information about other clues scientists use to study evolution. 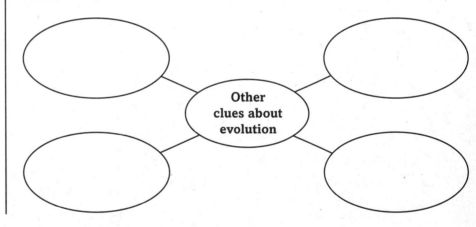

SYNTHESIZE IT A scientist discovers a new species of mammal. How could the scientist determine its evolutionary relationships to other animals? Explain how the scientist could use each type of evidence discussed in the section.

_____
_____
_____
_____

Adaptations over Time 69

Name _____ Date _____

Adaptations over Time
Section 3 The Evolution of Primates

Skim *Section 3 of your book. Read the headings. Write three questions that come to mind.*

1. _____
2. _____
3. _____

Review Vocabulary **Define** opposable *using your book.*

opposable _____
_____

New Vocabulary *Use your book to define the following terms. Then use each term in a sentence.*

primates _____
_____
_____
_____

hominid _____
_____
_____
_____

Homo sapiens _____
_____
_____
_____

Academic Vocabulary *Use a dictionary to define* similar.

similar _____
_____

Name _____ Date _____

Section 3 The Evolution of Primates (continued)

Main Idea	**Details**

Primates

I found this information on page _____.

Analyze adaptations that are common among primates by completing the table below. List three primate adaptations and the functions each allows.

Adaptation	Function

I found this information on page _____.

Distinguish three characteristics of hominids.

1. _____

2. _____

3. _____

I found this information on page _____.

Sequence the ancestors of early humans. Create a timeline of hominids in the boxes below. Identify and describe the hominid that lived during each time period.

Time period: 4–6 million years ago
Hominid:
Characteristics:

Time period: 1.5–2 million years ago
Hominid:
Characteristics:

Time period: 1.6 million years ago
Hominid:
Characteristics:

Adaptations over Time 71

Name _____ Date _____

Section 3 The Evolution of Primates (continued)

Main Idea | **Details**

Humans
I found this information on page _____.

Organize information about the origins of modern humans. Complete the diagram.

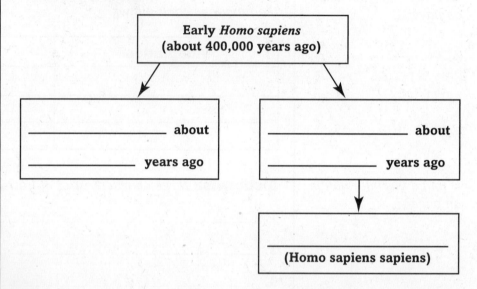

I found this information on page _____.

Contrast Neanderthals and Cro-Magnon humans by completing the diagram.

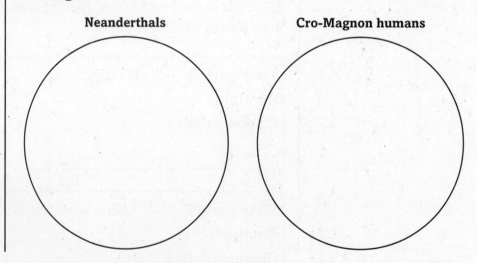

CONNECT IT Hypothesize how scientists might determine whether Neanderthals are ancestors of modern humans.

_____
_____
_____

72 Adaptations over Time

Tie It Together

Make Fossils

With a partner, model a set of fossils that show how organisms can change over time. Draw or model three related organisms. One should be the original organism. The others should be descendants of the original organism. Record the adaptations shown by your fossils. What environmental changes might have led to the adaptations?

Trade fossils with another pair. Describe the fossils that you are given. What adaptations can you find?

Adaptations over Time

Name _____ Date _____

Adaptations over Time Chapter Wrap-Up

Now that you have read the chapter, think about what you have learned and complete the table below. Compare your previous answers with these.

1. Write an **A** if you agree with the statement.
2. Write a **D** if you disagree with the statement.

Adaptations over Time	After You Read
• Traits acquired by an organism during its life can be passed on to its offspring.	
• Most evidence of evolution comes from fossils.	
• Organisms with traits best suited to their environment are more likely to survive and reproduce.	
• Humans share a common ancestor with other primates.	

Review

Use this checklist to help you study.

☐ Review the information you included in your Foldable.
☐ Study your *Science Notebook* on this chapter.
☐ Study the definitions of vocabulary words.
☐ Review daily homework assignments.
☐ Re-read the chapter and review the charts, graphs, and illustrations.
☐ Review the Self Check at the end of each section.
☐ Look over the Chapter Review at the end of the chapter.

SUMMARIZE IT After reading this chapter, identify three things that you have learned about adaptations of organisms over time.

_____
_____
_____
_____

Name _____ Date _____

Bacteria

Before You Read

Preview the chapter and section titles and the section headings. Complete the first two columns of the table by listing at least two ideas for each section in each column.

K What I know	W What I want to find out

Construct the Foldable as directed at the beginning of this chapter.

Science Journal

List ways that bacteria can be harmful and ways that bacteria can be beneficial. Which list is longer?

_____
_____
_____
_____
_____
_____

Bacteria 75

Name _____ Date _____

Bacteria

Section 1 What are bacteria?

Scan *Section 1 of the chapter.*

- Read all headings and bold words.
- Look at all of the illustrations.
- Think about what you already know about bacteria.

Write three facts that you learned while scanning the section.

1. _____
2. _____
3. _____

Define prokaryotic *to show its scientific meaning.*

prokaryotic _____
_____

Read the definitions below. Write the key term on the blank in the left column.

_____ organism that uses oxygen for respiration

_____ simplest form of asexual reproduction, in which two new cells are produced that have genetic material that is identical to each other and to the original cell

_____ whiplike tails that help many bacteria move

_____ organism that is adapted to live without oxygen

Use a dictionary to define the term **environment**.

environment _____
_____

Name _____ Date _____

Section 1 What are bacteria? (continued)

Main Idea — Details

Characteristics of Bacteria

I found this information on page _____.

Identify *3 shapes of bacterial cells.*

1. cocci: _____

2. bacilli: _____

3. spirilla: _____

I found this information on page _____.

Summarize *how the following pairs of words relate to* **bacteria.**

Asexual Reproduction/Sexual Reproduction: _____

_____

_____

_____

_____

_____

Producers/Consumers: _____

_____

_____

_____

_____

Aerobes/Anaerobes: _____

_____

_____

_____

_____

_____

Name _____ Date _____

Section 1 What are bacteria? (continued)

Main Idea	Details
Eubacteria *I found this information on page _____.*	**Complete** *the graphic organizer about the characteristics of* cyanobacteria. - produce _____ as waste - contain chlorophyll and _____ - **Cyanobacteria** - use carbon dioxide, sunlight, and water to _____ - are food for _____
I found this information on page _____.	**Summarize** *the different types of* consumer eubacteria. _____ _____ _____ _____ _____ _____
Archaebacteria *I found this information on page _____.*	**Identify** *three types of extreme environments in which* archaebacteria *can survive.* _____ _____
I found this information on page _____.	**Summarize** *how methane-producing bacteria obtain energy.* _____ _____

78 *Bacteria*

Name _____ Date _____

Bacteria

Section 2 Bacteria in Your Life

Skim the headings in Section 2. What do you think are two major ideas that will be discussed in this section?

1. _____

2. _____

Review Vocabulary **Define** disease *and use it in an original sentence.*

disease _____

_____

_____

New Vocabulary *Match the definitions with the appropriate key terms.*

_____ chemical produced by some bacteria that is used to limit the growth of other bacteria

_____ organism that uses dead organisms for food and energy

_____ bacteria that change nitrogen from the air into forms that plants and animals can use

_____ organism that causes disease

_____ poisonous substance produced by some pathogens

_____ thick-walled, protective structure produced by some bacteria when conditions are unfavorable for survival

_____ preparation made from killed bacteria or damaged particles from bacterial cell walls that can prevent some bacterial diseases

Academic Vocabulary *Use a dictionary to define the term* benefit.

benefit _____

_____

Bacteria 79

Name _____ Date _____

Section 2 Bacteria in Your Life (continued)

Main Idea | Details

Beneficial Bacteria

I found this information on page _____.

Analyze how some bacteria help you. Complete the paragraph.

_____ are helpful in many ways. Without them, you would not be able to stay _____ for very long. Bacteria in the _____ produce _____ which is needed for blood clotting. Some bacteria produce _____. These chemicals _____ of other bacteria.

I found this information on page _____.

Summarize the roles of *saprophytes* and *nitrogen-fixing bacteria* in the environment.

Role of saprophytes: _____
_____
_____

Nitrogen-fixing bacteria: _____
_____
_____

I found this information on page _____.

Complete the table describing some of the ways people use bacteria.

Human Uses for Bacteria	
Use	How do the bacteria help?
Bioremediation	
Food Production	
Industry	

80 *Bacteria*

Name _____ Date _____

Section 2 **Bacteria in Your Life** (continued)

Main Idea | Details

Harmful Bacteria

I found this information on page _____.

Analyze how pathogens *make you sick.* Complete the paragraph.

Pathogens can enter your body when you _____ and through _____. Once inside the body, they can multiply, _____, and cause _____.

I found this information on page _____.

Complete the graphic organizer about pasteurization.

| Most harmful bacteria are killed because _____. |
| The _____ _____ does not change. |

↓ Pasteurization ↓

| The process is used to prepare these foods: _____. |

I found this information on page _____.

Summarize information about vaccines.

_____
_____
_____
_____
_____

SUMMARIZE IT — Explain why it is important to learn about bacteria.

_____
_____
_____

Bacteria 81

Name _____ Date _____

Bacteria Chapter Wrap-Up

Review the ideas that you listed in the table at the beginning of the chapter. Cross out any incorrect information in the first column. Then complete the table by filling in the third column. How do your ideas about what you know now compare with those you provided at the beginning of the chapter?

K What I know	W What I want to find out	L What I learned

Review
Use this checklist to help you study.

☐ Review the information you included in your Foldable.
☐ Study your *Science Notebook* on this chapter.
☐ Study the definitions of vocabulary words.
☐ Review daily homework assignments.
☐ Re-read the chapter and review the charts, graphs, and illustrations.
☐ Review the Self Check at the end of each section.
☐ Look over the Chapter Review at the end of the chapter.

SUMMARIZE IT Identify three important ideas in this chapter.

_____
_____
_____
_____

Name _____ Date _____

Protists and Fungi

Before You Read

Before you read the chapter, respond to these statements.

1. Write an **A** if you agree with the statement.
2. Write a **D** if you disagree with the statement.

Before You Read	**Protists and Fungi**
	• Some protists have roots like those of plants.
	• The oxygen you breathe comes partly from green algae.
	• Protozoans are usually classified by what they eat.
	• Lichens can indicate the pollution level in an area.

 Construct the Foldable as directed at the beginning of this chapter.

Science Journal

In what ways might fungi benefit other organisms and the environment?

_____
_____
_____
_____
_____
_____

Protists and Fungi   83

Name _____ Date _____

Protists and Fungi
Section 1 Protists

Preview the What You'll Learn *statements for Section 1. Rewrite each statement as a question. Look for the answers as you read the section.*

1. _____
2. _____
3. _____
4. _____

Review Vocabulary **Define** asexual reproduction *to show its scientific meaning.*

asexual reproduction _____

New Vocabulary *Write the vocabulary word that matches each definition.*

_____ one-celled or many-celled eukaryotic organism that lives in moist or wet surroundings

_____ plantlike protists

_____ one-celled, animal-like protist

_____ long, thin, whiplike structure used for movement

_____ short, threadlike structures that extend from the cell membrane and help the organism move quickly

_____ temporary extension of cytoplasm that helps some protists move

Academic Vocabulary *Use a dictionary to define* visible.

visible _____

84 Protists and Fungi

Name _____ Date _____

Section 1 Protists (continued)

Main Idea | Details

What is a protist?

I found this information on page _____.

Compare and contrast the 3 groups of protists.

	Plantlike	Animal-like	Funguslike
Do they make their own food?			
Is there a cell wall?			
Can they move?			

Plantlike Protists

I found this information on page _____.

Summarize key information about plantlike protists.

Diatoms: _____
_____

Dinoflagellates: _____
_____

Euglenoids: _____
_____

Red algae: _____
_____

Green algae: _____
_____

Brown algae: _____
_____

Importance of Algae

I found this information on page _____.

Evaluate the importance of algae.

Algae in the Environment	Human Uses of Algae

Protists and Fungi 85

Name _____   Date _____

Section 1 Protists (continued)

Main Idea

Animal-Like Protists

I found this information on page _____.

Details

Classify protozoans. *Summarize key information about each type of protozoan.*

Type	Characteristics

Importance of Protozoans

I found this information on page _____.

Summarize the importance of protozoans to other organisms.

_____

_____

_____

Funguslike Protists and Importance of Funguslike Protists

I found this information on page _____.

Complete the prompts with information about funguslike protists.

Funguslike protists produce _____ like fungi and must take in food from _____. Slime molds use _____ to move and live on _____ or _____ in moist, cool, shady environments. Downy molds and mildews grow as a mass of _____ over an organism. Some are parasites; others feed on _____. Funguslike protists in the ecosystem help break down _____. Some are _____ to other organisms.

CONNECT IT Why is it dangerous to drink water from unknown sources?

_____

_____

_____

86 *Protists and Fungi*

Name _____ Date _____

Protists and Fungi
Section 2 Fungi

Skim Section 2. Predict two topics that will be covered.

1. _____

2. _____

Review Vocabulary **Define** photosynthesis *using your book or a dictionary.*

photosynthesis _____

_____

_____

New Vocabulary Write the correct vocabulary word next to its definition.

_____ mass of threadlike tubes forming the body of a fungus

_____ organism that absorbs energy from dead and decaying tissues

_____ waterproof reproductive cell that can grow into a new organism

_____ reproductive cells produced by club fungi

_____ reproductive cells produced by sac fungi

_____ form of asexual reproduction in which a new, genetically identical organism forms on the side of its parent

_____ case containing reproductive cells produced by some types of fungi

_____ organism made up of a fungus and a green alga or a cyanobacterium

_____ network of hyphae and plant roots that helps plants absorb water and minerals from the soil

Academic Vocabulary Use a dictionary to define decline.

decline _____

_____

Name _____ Date _____

Section 2 Fungi (continued)

Main Idea

What are fungi?

I found this information on page _____.

Details

Complete *the table to describe the characteristics of* fungi.

Structure	Obtaining Food
Reproduction	Differences from Plants

Club Fungi, Sac Fungi, Zygote Fungi, and Other Fungi

I found this information on page _____.

Compare *club, sac, and zygote fungi.*

	Examples	How they reproduce
Club fungi		
Sac fungi		
Zygote fungi		

I found this information on page _____.

Summarize *why some fungi are difficult to classify.*

_____

_____

Name _____ Date _____

Section 2 Fungi (continued)

Main Idea — **Details**

Lichens, Fungi, and Plants

I found this information on page _____.

Identify three important roles of lichens.

1. _____
2. _____
3. _____

I found this information on page _____.

Model the beneficial relationship between fungi and plants by completing the diagram.

Some fungi and plants form a network of _____ and _____ called _____.

| The fungi help the plants _____. | ↔ | The plants supply _____ and _____ to the fungi. |

The Importance of Fungi

I found this information on page _____.

Identify the importance of fungi in each of these areas.

Foods	Agriculture	Health and Medicine	Decomposers

CONNECT IT Describe what nature would be like without lichens, mycorrhizae, and decomposer fungi.

_____
_____
_____

Protists and Fungi 89

Name _____   Date _____

Protists and Fungi   Chapter Wrap-Up

Now that you have read the chapter, think about what you have learned and complete the table below. Compare your previous answers with these.

1. Write an **A** if you agree with the statement.
2. Write a **D** if you disagree with the statement.

Protists and Fungi	**After You Read**
• Some protists have roots like those of plants.	
• The oxygen you breathe comes partly from green algae.	
• Protozoans are usually classified by what they eat.	
• Lichens can indicate the pollution level in an area.	

Review

Use this checklist to help you study.

☐ Review the information you included in your Foldable.
☐ Study your *Science Notebook* on this chapter.
☐ Study the definitions of vocabulary words.
☐ Review daily homework assignments.
☐ Re-read the chapter and review the charts, graphs, and illustrations.
☐ Review the Self Check at the end of each section.
☐ Look over the Chapter Review at the end of the chapter.

SUMMARIZE IT   After reading the chapter, write three facts you learned that you did not know before.

_____
_____
_____
_____

Name _____ Date _____

Plants

Before You Read

Before you read the chapter, respond to these statements.

1. Write an **A** if you agree with the statement.
2. Write a **D** if you disagree with the statement.

Before You Read	**Plants**
	• In tropical rain forests, there are more than 260,000 known plant species and probably more to be identified.
	• Land plants' ancestors may have been green algae that lived in the sea.
	• Ferns and mosses produce spores rather than seeds.
	• Paper and clothing are made from seed plants.

 Construct the Foldable as directed at the beginning of this chapter.

Science Journal
Write three characteristics that you think all plants have in common.

_____
_____
_____
_____
_____
_____
_____

Plants 91

Plants

Section 1 An Overview of Plants

Skim the headings in Section 1. Then predict three facts you will learn from reading the section.

1. _____
2. _____
3. _____

Review Vocabulary **Define** the word species. Use your book or a dictionary for help.

species _____

New Vocabulary Use your book to define the following key terms.

cuticle _____

cellulose _____

vascular plant _____

nonvascular plant _____

Academic Vocabulary Use a dictionary to define adapt to reflect its scientific meaning.

adapt _____

Section 1 An Overview of Plants (continued)

Main Idea

What is a plant?

I found this information on page _____.

Details

Summarize how plants make food by completing the concept map below. Use these terms: photosynthesis, chlorophyll, chloroplasts.

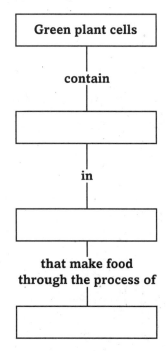

Main Idea

Origin and Evolution of Plants

I found this information on page _____.

Sequence the events in the table below. Write the oldest event at the bottom of the table and the youngest event at the top of the table.

Events
- First cone-bearing plants
- First green algae
- First flowering plants
- First land plants

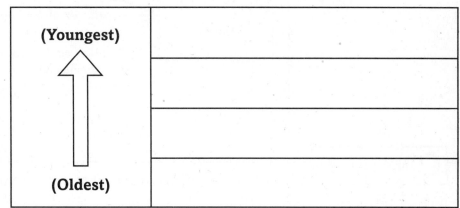

Plants 93

Name _____ Date _____

Section 1 An Overview of Plants (continued)

Main Idea — **Details**

Life on Land
I found this information on page _____.

Summarize how land plants made life possible for land animals.
_____
_____
_____

Adaptations to Land
I found this information on page _____.

Identify the four adaptations that make it possible for plants to live on land.

Plant Adaptations to Land	
Structure	Function

Classification of Plants
I found this information on page _____.

Complete the concept map below about plant classification.

Classification of plants
- divides plants into two major groups called → ☐
- was developed by → ☐
- gives each plant species its own → ☐

CONNECT IT Suppose that you are working at a greenhouse. While at work, a child asks you, "What's a plant?" Write a short answer to this question.
_____
_____

Plants

Section 2 Seedless Plants

Skim *Section 2 of your book. Then write three questions that you have about plants. Try to answer your questions as you read.*

1. _____
2. _____
3. _____

Review Vocabulary

Define spore. *Use your book or a dictionary for help. Write a sentence that reflects its scientific meaning.*

spore _____

New Vocabulary

Use your book to define the following key terms. Then use each word in a sentence that reflects its scientific meaning.

rhizoid _____

pioneer species _____

Academic Vocabulary

Use a dictionary to define soil. *Write a sentence that reflects its scientific meaning.*

soil _____

Plants 95

Name _____   Date _____

Section 2 Seedless Plants (continued)

Main Idea

Seedless Nonvascular Plants

I found this information on page _____.

Details

Organize the characteristics of seedless nonvascular plants by completing the chart below.

Characteristics of Seedless Nonvascular Plants
1.
2.
3.
4.
5.
6.
7.
8.

I found this information on page _____.

Complete the concept map to identify examples and characteristics of seedless nonvascular plants. One example has been listed for you.

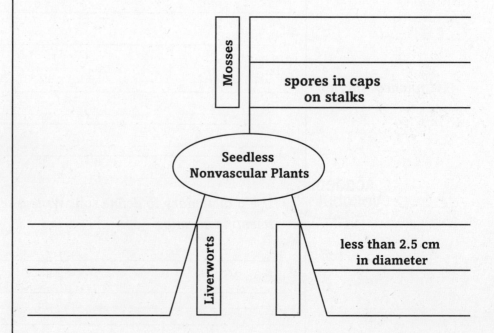

96 Plants

Name _____ Date _____

Section 2 Seedless Plants (continued)

Main Idea | Details

Seedless Vascular Plants

I found this information on page _____.

Compare and contrast seedless vascular plants *with* seedless nonvascular plants *in the Venn diagram below.*

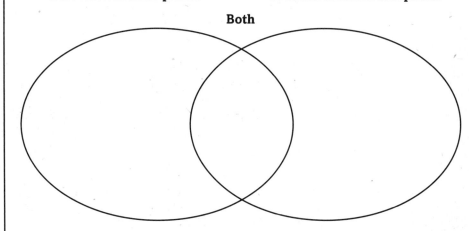

Importance of Seedless Plants

I found this information on page _____.

Summarize *the importance of seedless plants in the table below.*

Importance of Seedless Plants
1.
2.
3.
4.
5.
6.
7.

CONNECT IT Suppose you are a naturalist working in a forest area that has recently burned in a forest fire. Summarize what you would tell visitors about seedless plants and how important they are to the forest's recovery.

_____
_____
_____

Plants 97

Name _____   Date _____

Plants
Section 3 Seed Plants

Scan *Section 3 of your book. Write three questions that come to mind as you read the headings and examine the illustrations.*

1. _____
2. _____
3. _____

Review Vocabulary **Define** *seed. Use your book or a dictionary for help. Then use this word in a sentence that reflects its scientific meaning.*

seed _____
_____

New Vocabulary *Read the definitions below. Write the correct key term on the blank in the left column. Use your book for help.*

_____ a vascular plant that produces seeds that are not protected by fruit

_____ a vascular plant that flowers and produces fruit with one or more seeds

_____ a plant with one cotyledon inside its seeds

_____ a plant with two cotyledons inside its seeds

Academic Vocabulary *Use a dictionary to define* annual *as it applies to the length of a plant's life.*

annual _____
_____

Name _____ Date _____

Section 3 **Seed Plants** (continued)

Main Idea	**Details**

Characteristics of Seed Plants

I found this information on page _____ .

Create *a cross-section of a* leaf *in the space below. Label and describe the purpose of six important features.*

I found this information on page _____ .

Organize *the characteristics of* seed plants *by completing the chart below.*

Structure	Function
Leaves	
Stems	
Roots	
Vascular tissue	

Plants 99

Name _____ Date _____

Section 3 Seed Plants (continued)

Main Idea — Details

Gymnosperms

I found this information on page _____.

Complete the chart below about gymnosperms by writing about the characteristic listed in that cell.

Gymnosperms	
Divisions	Seeds
Flowers	Leaves

Angiosperms

I found this information on page _____.

Complete the chart below about angiosperms by writing about the characteristic listed in that cell.

Angiosperms	
Division	Seeds
Flowers	Fruits

Importance of Seed Plants

I found this information on page _____.

Skim your book for two uses each of gymnosperms and angiosperms.

Gymnosperms:

1. _____

2. _____

Angiosperms:

1. _____

2. _____

100 *Plants*

Name _____ **Date** _____

Tie It Together

In the space below, draw a sketch of a tree. Label the tree's roots, trunk, and leaves. Next to each label, write the important functions that each of these structures performs. Beneath your sketch, explain why trees are an important part of the environment.

Name _____ Date _____

Plants Chapter Wrap-Up

Now that you have read the chapter, think about what you have learned and complete the table below. Compare your previous answers with these.

1. Write an **A** if you agree with the statement.
2. Write a **D** if you disagree with the statement.

Plants	After You Read
• In tropical rain forests, there are more than 260,000 known plant species and probably more to be identified.	
• Land plants' ancestors may have been green algae that lived in the sea.	
• Ferns and mosses produce spores rather than seeds.	
• Paper and clothing are made from seed plants.	

Review

Use this checklist to help you study.

☐ Review the information you included in your Foldable.
☐ Study your *Science Notebook* on this chapter.
☐ Study the definitions of vocabulary words.
☐ Review daily homework assignments.
☐ Re-read the chapter and review the charts, graphs, and illustrations.
☐ Review the Self Check at the end of each section.
☐ Look over the Chapter Review at the end of the chapter.

SUMMARIZE IT After reading this chapter, identify three things that you have learned about plants.

_____
_____
_____
_____

Plant Reproduction

Before You Read

Before you read the chapter, respond to these statements.

1. Write an **A** if you agree with the statement.
2. Write a **D** if you disagree with the statement.

Before You Read	Plant Reproduction
	• Both humans and plants need water, oxygen, energy, and food to grow.
	• Ferns and mosses reproduce by forming spores.
	• All seeds are produced by flowering plants.
	• Some seeds are spread by gravity.

 Construct the Foldable as directed at the beginning of this chapter.

Science Journal

List three plants that reproduce by forming seeds.

_____
_____
_____
_____
_____
_____
_____

Name _____ Date _____

Plant Reproduction
Section 1 Introduction to Plant Reproduction

Scan *Section 1 of your book using the checklist below.*

☐ Read all section titles.
☐ Read all bold words.
☐ Read all charts and graphs.
☐ Look at all the pictures and read their captions.
☐ Think about what you already know about plant reproduction.

Write three facts that you discovered about plant reproduction as you scanned this section.

1. _____
2. _____
3. _____

Review Vocabulary **Define** fertilization *in a sentence that shows its scientific meaning.*

fertilization _____

New Vocabulary *Use your book to define the following terms.*

spore _____
_____

gametophyte stage _____
_____

sporophyte stage _____
_____

Academic Vocabulary *Use a dictionary to define* identical.

identical _____

104 Plant Reproduction

Name _____ Date _____

Section 1 Introduction to Plant Reproduction (continued)

Main Idea | **Details**

Types of Reproduction

I found this information on page _____.

Compare and contrast *two ways that plants reproduce.*

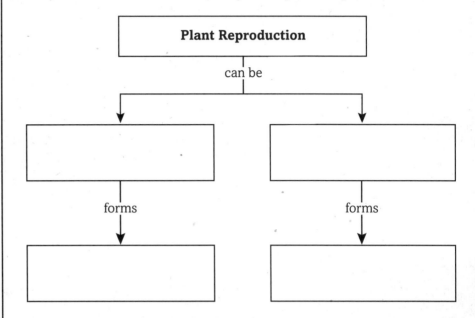

I found this information on page _____.

Sequence *the steps in plant fertilization. Complete the flow chart.*

| Female reproductive structures produce _____. | Male reproductive structures produce _____. |

Are both structures found on the same plant?

No → [ ]   Yes → [ ]

Plant Reproduction 105

Section 1 Introduction to Plant Reproduction (continued)

Main Idea

Plant Life Cycles

I found this information on page _____.

Details

Model *the two stages of a plant's life cycle by labeling the diagram below with the following terms.*

- gametophyte plant structures (*n*)
- sex cells (sperm and eggs) (*n*)
- sporophyte plant structures (*2n*)
- spores (*n*)

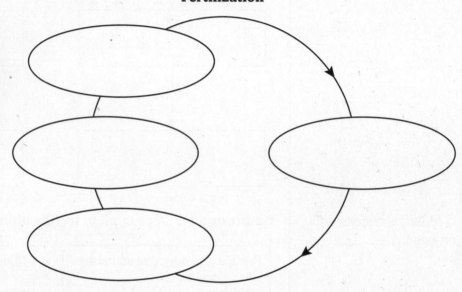

Contrast *the gametophyte and sporophyte stages of plant development. Complete the table.*

Stage	Cell type	Reproductive cells formed	How reproductive cells form
Gametophyte			
Sporophyte			

CONNECT IT A plant breeder wants to develop new varieties of roses that have different traits from the varieties he already has. Describe the type of reproduction the breeder is most likely to use and why.

_____
_____
_____

106 Plant Reproduction

Plant Reproduction
Section 2 Seedless Reproduction

Skim *Section 2 of your book. Read the headings and look at the illustrations. Write three questions that come to mind.*

1. _____
2. _____
3. _____

Review Vocabulary
Define photosynthesis *using your book or a dictionary.*

photosynthesis _____

New Vocabulary
Use your book to define the following terms.

frond _____

rhizome _____

sori _____

prothallus _____

Academic Vocabulary
Use a dictionary to define widespread.

widespread _____

Name _____ Date _____

Section 2 Seedless Reproduction (continued)

Main Idea	Details
The Importance of Spores *I found this information on page _____.*	**Summarize** *the role of spores in plant reproduction.* Spores are used by _____ _____ to reproduce. The _____ stage of the plant produces _____ spores in _____. These _____, and the spores are spread by _____. The spores grow into _____ that can produce _____.
Nonvascular Seedless Plants *I found this information on page _____.*	**Sequence** *the life cycle of a moss. Complete the flow chart.* • _____ begins the sporophyte stage. • _____ occurs, producing _____ spores. • Spores grow into _____ _____ that produce _____. • _____ swims to the _____, and _____ occurs.
I found this information on page _____.	**Distinguish** *two ways in which nonvascular plants reproduce asexually.* 	Type of Plant	Asexual Reproduction Process	 |---|---| 	moss		 	liverwort		

108   Plant Reproduction

Name _____ Date _____

Section 2 Seedless Reproduction (continued)

Main Idea	**Details**

Vascular Seedless Plants

I found this information on page _____.

Contrast *vascular and nonvascular seedless plants. Complete the Venn diagram with at least six facts.*

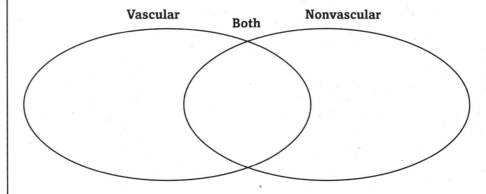

I found this information on page _____.

Organize *the life cycle of a fern into a flow chart.*

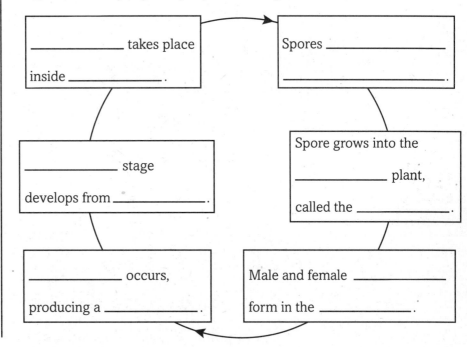

CONNECT IT Suppose that you are walking through a forest and you see some moss plants and ferns. Describe how you could know the stage of its life cycle each kind of plant is in.

_____
_____
_____

Name _____ Date _____

Plant Reproduction
Section 3 Seed Reproduction

Predict three things that will be discussed in Section 3.

1. _____
2. _____
3. _____

Review Vocabulary **Define** gymnosperms *using your book or a dictionary.*

gymnosperms _____
_____
_____
_____

New Vocabulary *Match each vocabulary term to its definition.*

_____ small structure produced by the male reproductive organs of a seed plant

_____ transfer of pollen grains to the female part of a seed plant

_____ series of events that results in the growth of a plant from a seed

_____ part of a plant that produces the egg

_____ male reproductive organ in a flower

_____ female reproductive organ in a flower

_____ part of a flower in which ovules are found

Academic Vocabulary *Use a dictionary to define* structure *as it is used in science.*

structure _____
_____
_____

110 Plant Reproduction

Name _____ Date _____

Section 3 Seed Reproduction (continued)

Main Idea	Details

The Importance of Pollen and Seeds
 	
I found this information on page _____.
	Summarize *key facts about* pollen *and* pollination. *Complete the outline.*
	Pollen and Pollination in Seed Plants
	  I. Pollen grains
	    A. _____
	    B. _____
	  II. Pollination
	    A. _____
	        _____
	    B. _____
	        _____

I found this information on page _____.	**Model** *a seed. Draw a seed and label the* stored food, embryo, *and* seed coat. *Identify the role of each part of the seed.*
Gymnosperm Reproduction *I found this information on page _____.*	**Sequence** *steps of* gymnosperm seed formation *in the flow chart.* Male: _____ produced in _____ Female: _____ produced in _____ in _____ carried by _____ fertilization

Plant Reproduction 111

Name _____ Date _____

Section 3 Seed Reproduction (continued)

Main Idea	**Details**

Angiosperm Reproduction

I found this information on page _____.

Model *a flower by drawing and labeling its parts. Then write a brief caption to identify the male and female reproductive organs and to describe how each organ functions during fertilization.*

[blank box for drawing]

Seed Dispersal

I found this information on page _____.

Sequence *the events of* fertilization *and* germination *in angiosperms.*

1. Flower is _____.
2. _____.
3. _____.
4. Seed is _____.
5. Conditions become right for _____.
6. _____.
7. _____.
8. Root grows from _____.
9. _____.
10. Photosynthesis begins.

CONNECT IT The seeds of horse chestnut trees are covered with a prickly outer layer. Propose a way that you think these seeds might be dispersed.

_____
_____

112 *Plant Reproduction*

Tie It Together

Describe a Plant

Suppose that you are an explorer who has discovered a new species of plant.
- *Draw and describe the plant below.*
- *Be sure to indicate whether your plant is vascular or nonvascular.*
- *If it does reproduce with seeds, identify it as an angiosperm or a gymnosperm.*
- *Include a diagram that shows the plant's life cycle.*
- *Draw a cross-section of the plant that identifies its reproductive structures.*

Name _____ Date _____

Plant Reproduction Chapter Wrap-Up

Now that you have read the chapter, think about what you have learned and complete the table below. Compare your previous answers with these.

1. Write an **A** if you agree with the statement.
2. Write a **D** if you disagree with the statement.

Plant Reproduction	**After You Read**
• Both humans and plants need water, oxygen, energy, and food to grow.	
• Ferns and mosses reproduce by forming spores.	
• All seeds are produced by flowering plants.	
• Some seeds are spread by gravity.	

Review

Use this checklist to help you study.

- ☐ Review the information you included in your Foldable.
- ☐ Study your *Science Notebook* on this chapter.
- ☐ Study the definitions of vocabulary words.
- ☐ Review daily homework assignments.
- ☐ Re-read the chapter and review the charts, graphs, and illustrations.
- ☐ Review the Self Check at the end of each section.
- ☐ Look over the Chapter Review at the end of the chapter.

SUMMARIZE IT After reading this chapter, identify three things that you have learned about plant reproduction.

_____
_____
_____
_____

Name _____ Date _____

Plant Processes

Before You Read

Before you read the chapter, respond to these statements.

1. Write an **A** if you agree with the statement.
2. Write a **D** if you disagree with the statement.

Before You Read	**Plant Processes**
	• Plants make their own food.
	• Plants break down food to release energy.
	• Plant stems grow away from light.
	• Plants have hormones that control changes in their growth.

 Construct the Foldable as directed at the beginning of this chapter.

Science Journal

Describe what would happen to life on Earth if all the green plants disappeared.

_____
_____
_____
_____
_____
_____
_____

Name _____ Date _____

Plant Processes
Section 1 Photosynthesis and Respiration

Scan the illustrations in Section 1. Write three questions that you have about plants. Try to answer your questions as you read.

1. _____
2. _____
3. _____

Review Vocabulary **Define** cellulose *using your book. Then write a sentence to illustrate its scientific meaning.*

cellulose _____
_____
_____
_____

New Vocabulary Use your book to define the following terms.

stomata _____
_____

chlorophyll _____
_____

photosynthesis _____
_____

respiration _____
_____

Academic Vocabulary *Use a dictionary to define* release.

release _____

116 *Plant Processes*

Name _____ Date _____

Section 1 Photosynthesis and Respiration (continued)

Main Idea | Details

Taking In Raw Materials

I found this information on page _____.

Organize *what you know about the different layers of a plant's leaves by completing the table below.*

Structure	Function
Epidermis	
Palisade layer	
Spongy layer	

I found this information on page _____.

Summarize *why stomata are important structures in a plant leaf.*

_____
_____
_____

The Food-Making Process

I found this information on page _____.

Complete *the equation for* photosynthesis. *Identify:*

- the product that is stored as a food source
- the product that is released mostly as waste
- the product made during light-dependent reactions
- the product made during light-independent reactions

$$6CO_2 + 6H_2O + \text{light energy} \longrightarrow \underline{} + \underline{}$$

carbon dioxide | water | Food source: _____ made during _____ | Waste product: _____ made during _____

Plant Processes 117

Name _____ Date _____

Section 1 Photosynthesis and Respiration (continued)

Main Idea	**Details**

The Breakdown of Food

I found this information on page _____.

Define aerobic respiration.

_____

Complete the equation for aerobic respiration.

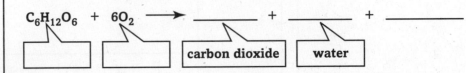

$$C_6H_{12}O_6 + 6O_2 \longrightarrow \underline{} + \underline{} + \underline{}$$

(carbon dioxide) (water)

Comparison of Photosynthesis and Respiration

I found this information on page _____.

Compare the processes of photosynthesis and aerobic respiration by completing the table.

	Photosynthesis	Aerobic Respiration
Energy		
Raw materials		
End products		
Cell structure in which process occurs		

SUMMARIZE IT Create a concept map or other diagram to summarize what you learned in this section about plant structure and function.

118 Plant Processes

Plant Processes
Section 2 Plant Responses

Scan Section 2. Predict three things that you will learn.

1. _____
2. _____
3. _____

Review Vocabulary **Define** behavior *using your book.*

behavior _____
_____

New Vocabulary Write the correct vocabulary term next to each definition. Use your book to help you.

_____ response of a plant to external stimuli, movement caused by change in growth

_____ type of plant hormone that causes plant stems and leaves to exhibit positive responses to light

_____ plant's response to the number of hours of daylight and darkness it receives

_____ plant that generally requires short nights—less than 12 hours of darkness—to begin the flowering process

_____ plant that generally requires long nights—12 or more hours of darkness—to begin the flowering process

_____ plant that does not require a specific photoperiod and can begin the flowering process over a range of night lengths

Academic Vocabulary *Use a dictionary to define* involve.

involve _____
_____

Plant Processes 119

Name _____ Date _____

Section 2 Plant Responses (continued)

Main Idea	Details

What are plant responses?

I found this information on page _____.

Distinguish the types of stimuli as internal or external.

_____ 1. a stimulus that comes from outside the body

_____ 2. a stimulus that comes from inside the body

Tropisms

I found this information on page _____.

Complete the table below. Identify the stimulus for each described response.

Stimulus	Response
	Plant stem grows faster on one side. Stem bends and twists around object.
	Plant bends toward light. Leaves turn and absorb more light.
	Roots grow downward. Stems grow upward.

Plant Hormones

I found this information on page _____.

Compare the effects of different hormones that affect plants.

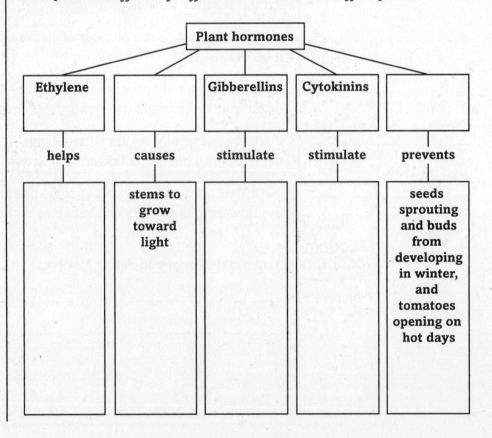

120 *Plant Processes*

Name _____ Date _____

Section 2 Plant Responses (continued)

Main Idea | Details

Plant Hormones

I found this information on page _____.

Create *a diagram to illustrate how auxin causes a stem to grow in response to sunlight. Write a short caption to describe where* auxin *is concentrated in the stem.*

Photoperiods

I found this information on page _____.

Complete *the table below to show your understanding of the effects of* photoperiodism *on different types of plants.*

Type of Plant	Hours of Darkness Needed to Flower	Examples
	need less than 12 hours	spinach, lettuce, and beets
	need 12 or more hours	poinsettias, strawberries, and ragweed
	do not need a specific amount of light	dandelions and roses

CONNECT IT Explain plant responses you might see in plants that are growing indoors on a windowsill.

_____
_____
_____

Plant Processes 121

Name _____ Date _____

Plant Processes Chapter Wrap-Up

Now that you have read the chapter, think about what you have learned and complete the table below. Compare your previous answers with these.

1. Write an **A** if you agree with the statement.
2. Write a **D** if you disagree with the statement.

Plant Processes	After You Read
• Plants make their own food.	
• Plants break down food to release energy.	
• Plant stems grow away from light.	
• Plants have hormones that control changes in their growth.	

Review
Use this checklist to help you study.

☐ Review the information you included in your Foldable.
☐ Study your *Science Notebook* on this chapter.
☐ Study the definitions of vocabulary words.
☐ Review daily homework assignments.
☐ Re-read the chapter and review the charts, graphs, and illustrations.
☐ Review the Self Check at the end of each section.
☐ Look over the Chapter Review at the end of the chapter.

SUMMARIZE IT After reading this chapter, identify three things that you have learned about plant processes.

_____
_____
_____
_____

Name _____ Date _____

Introduction to Animals

Before You Read

Before you read the chapter, think about what you know about the topic. List three things that you already know about animals in the first column. Then list three things that you would like to learn about animals in the second column.

K What I know	W What I want to find out

 Construct the Foldable as directed at the beginning of this chapter.

Science Journal

List the animals you may find living around a coral reef.

_____
_____
_____
_____
_____
_____
_____

Introduction to Animals   123

Name _____   Date _____

Introduction to Animals
Section 1 Is it an animal?

Scan the headings in Section 1 of the chapter. Identify three topics that are discussed.

1. _____
2. _____
3. _____

Review Vocabulary **Define** adaptation *using your book or a dictionary.*

adaptation
_____
_____

New Vocabulary Read the definitions below. Write the correct vocabulary term on the blank to the left of each definition.

_____ animal that eats both plants and animals; mammals with specialized teeth for eating plants and animals

_____ arrangement of body parts in a circle around a center point

_____ an animal without a backbone

_____ animal that eats only other animals or the remains of other animals

_____ arrangement of body parts into halves that are nearly mirror images of each other

_____ animal that eats only plants or parts of plants

_____ an animal that has a backbone

Academic Vocabulary Use a dictionary to define definite *to show its scientific meaning.*

definite
_____
_____

124 Introduction to Animals

Name _____ Date _____

Section 1 Is it an animal? (continued)

Main Idea

Animal Characteristics

I found this information on page _____.

Details

Summarize *the characteristics of animals by completing the following main points.*

Animals get their food from _____.

Many animals move from place to place to find _____, _____, and/or _____.

All animals can reproduce _____. Some also can reproduce _____.

Animal cells have a _____ and other parts inside called _____.

How Animals Meet Their Needs

I found this information on page _____.

Compare *animal adaptations by completing the chart.*

How Animals Meet Their Needs		
	Adaptations	Animal Examples
Ways to get energy	eat plants	deer, some fishes
Physical features	large size	moose, bison
Behaviors		

Introduction to Animals   125

Name _____ Date _____

Section 1 Is it an animal? (continued)

Main Idea | ## Details

Animal Classification

I found this information on page _____.

Complete *and label the circle graph to compare the percent of known animals that are vertebrates with the percent of known animals that are invertebrates.*

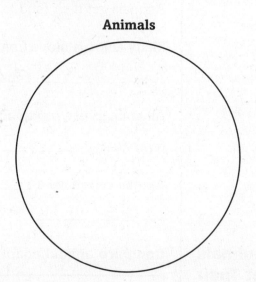

Animals

I found this information on page _____.

Compare *forms of animal symmetry by identifying and drawing an example of each below.*

Asymmetrical	Radial Symmetry	Bilateral Symmetry

SUMMARIZE IT Analyze the physical or behavioral adaptations of an animal that protect it from predators.

_____
_____
_____
_____

126 *Introduction to Animals*

Name _____ Date _____

Introduction to Animals
Section 2 Sponges and Cnidarians

Skim *Section 2 of the chapter. Read the headings and look at the illustrations. Predict three things that you will learn.*

1. _____
2. _____
3. _____

Review Vocabulary **Define** flagella *using your book or a dictionary.*

flagella _____
_____

New Vocabulary *Read the definitions below. Write the correct vocabulary term on the blank to the left of each definition.*

_____ form of a cnidarian that is bell-shaped and free-swimming

_____ capsule with a threadlike structure containing toxins that help a cnidarian capture food

_____ organisms that remain attached to one place during most of their life

_____ armlike structures that have stinging cells used for getting food

_____ animal that produces both sperm and eggs in the same body

_____ cnidarian body type that is vase-shaped and is usually sessile

Academic Vocabulary *Use a dictionary to define* source *to show its scientific meaning.*

source _____
_____

Introduction to Animals   127

Name _____ Date _____

Section 2 Sponges and Cnidarians (continued)

Main Idea	Details
Sponges and **Characteristics of Sponges** I found this information on page _____. I found this information on page _____.	**Summarize** *information about* sponges. Sponges appeared on Earth about _____. Most live in _____. Some have _____ symmetry, but most are _____. Adult sponges are _____, which means they do not move. Sponges pull _____ into their bodies, where cells filter out _____ and _____. **Model** *a sponge's body. Label the sponge's* central cavity *and* pores. *Show the* path *followed by water into and out of the sponge.*

[blank box for drawing]

Cnidarians I found this information on page _____.	**Organize** *information about the* two forms of cnidarians *by completing the chart.*

	Medusa	Polyp
Body Form (shape)		
Mobility		usually sessile
Examples	jellyfishes for most of their lives	

128 Introduction to Animals

Name _____ Date _____

Section 2 Sponges and Cnidarians (continued)

Main Idea	**Details**

Cnidarians

I found this information on page _____.

Sequence *the steps in reproduction of medusa forms of cnidarians by completing the cycle chart.*

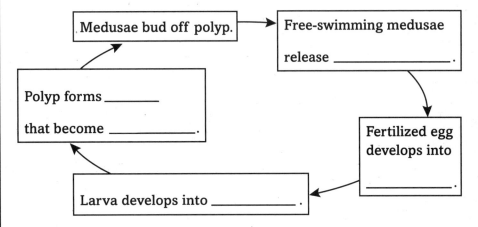

Corals

I found this information on page _____.

Summarize *key information about coral reefs in the outline.*

I. Coral reefs
   A. Formation of coral reefs
     1. Made of _____
     2. Grow as _____
     _____
     3. Can take _____ of years to form
   B. Importance of coral reefs
     1. Provide habitat for _____
     2. Protect _____
     3. Provide _____

SYNTHESIZE IT Explain how sponges and cnidarians could be mistaken for plants rather than animals.

_____
_____
_____
_____

Introduction to Animals 129

Name _____ Date _____

Introduction to Animals
Section 3 Flatworms and Roundworms

Scan *Section 3 of the chapter. Write four questions that come to mind. Look for answers to your questions as you read the section.*

1. _____
2. _____
3. _____
4. _____

Review Vocabulary   **Define** cilia *using your book or a dictionary.*

cilia _____
_____

New Vocabulary   *Use your book or a dictionary to define each vocabulary term. Then use each term in a sentence that shows its scientific meaning.*

free-living organisms _____
_____
_____
_____
_____
_____

anus _____
_____
_____
_____
_____

Academic Vocabulary   *Use a dictionary to define* require *to show its scientific meaning.*

require _____

Name _____ Date _____

Section 3 **Flatworms and Roundworms** (continued)

Main Idea | Details

What is a worm?

I found this information on page _____.

Analyze worms *by identifying four characteristics below.*

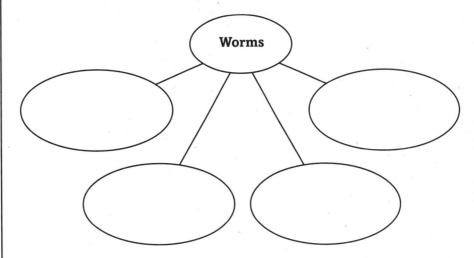

Flatworms

I found this information on page _____.

Compare characteristics of planarians and flukes *by completing the chart below.*

Flatworms		
	Planarians	**Flukes**
How they live		as parasites
What they eat		
How they move		
How they reproduce		usually sexually

I found this information on page _____.

Model a tapeworm *by sketching it. Label its hooks, its suckers, and a mature segment with eggs.*

Introduction to Animals 131

Name _____ Date _____

Section 3 **Flatworms and Roundworms** (continued)

Main Idea — **Details**

Origin of Flatworms

I found this information on page _____.

Summarize what some scientists believe about the *origin of flatworms* by completing the diagram.

probably the first group of animals to have ⟩ _____

Roundworms

I found this information on page _____.

Compare and contrast roundworms *with* flatworms *by completing the Venn diagram with at least seven facts.*

Flatworms   Both   Roundworms

CONNECT IT Summarize ways that roundworms are both helpful and harmful.

_____
_____
_____
_____

132 *Introduction to Animals*

Tie It Together

Preventing Disease

You are working on a public health campaign to inform people of the dangers of parasitic flatworms and roundworms. Create a poster with key information about diseases these organisms can cause and how to avoid them. Use words, pictures, and diagrams to get your message across.

Name _____ Date _____

Introduction to Animals Chapter Wrap-Up

Review the ideas you listed in the chart at the beginning of the chapter. Cross out any incorrect information in the first column. Then complete the chart by filling in the third column.

K What I know	W What I want to find out	L What I learned

Review
Use this checklist to help you study.

☐ Review the information you included in your Foldable.
☐ Study your *Science Notebook* on this chapter.
☐ Study the definitions of vocabulary words.
☐ Review daily homework assignments.
☐ Re-read the chapter and review the charts, graphs, and illustrations.
☐ Review the Self Check at the end of each section.
☐ Look over the Chapter Review at the end of the chapter.

SUMMARIZE IT After reading this chapter, identify three main ideas that you have learned about animals.

_____
_____
_____
_____

Name _____ Date _____

Mollusks, Worms, Arthropods, Echinoderms

Before You Read

Before you read the chapter, think about what you know about the topic. List three things you already know about mollusks, worms, arthropods, and echinoderms in the first column. Then list three things you would like to learn about them in the second column.

K What I know	W What I want to find out

   Construct the Foldable as directed at the beginning of this chapter.

Science Journal

List three animals from each animal group you will be studying: mollusks, worms, arthropods, and echinoderms.

_____
_____
_____
_____
_____

Mollusks, Worms, Arthropods, Echinoderms **135**

Name _____   Date _____

Mollusks, Worms, Arthropods, Echinoderms
Section 1 Mollusks

Scan the headings in Section 1 of your book. Identify three topics that will be discussed.

1. _____
2. _____
3. _____

Review Vocabulary **Define** visceral mass *using your book or a dictionary.*

visceral mass _____
_____

New Vocabulary *Use your book or a dictionary to define the following terms.*

mantle _____
_____
_____

gill _____
_____
_____

open circulatory system _____
_____
_____

radula _____
_____
_____

closed circulatory system _____
_____
_____

Academic Vocabulary *Use a dictionary to define* relax *as it might be used in science.*

relax _____

136 *Mollusks, Worms, Arthropods, Echinoderms*

Name _____ Date _____

Section 1 Mollusks (continued)

Main Idea | **Details**

Characteristics of Mollusks

I found this information on page _____ .

Identify characteristics of mollusks *in the chart below.*

Characteristics of Mollusks	
Type of symmetry	bilateral
Body description	
Where they live	

I found this information on page _____ .

Model *the* body of a mollusk *by sketching a snail and labeling its* shell, mantle, gill, mantle cavity, foot, radula, *and other body parts.*

Mollusks, Worms, Arthropods, Echinoderms

Name _____ Date _____

Section 1 Mollusks (continued)

Main Idea	**Details**

Classification of Mollusks

I found this information on page _____.

Compare and contrast types of mollusks *by completing the chart.*

Types of Mollusks			
Types	Gastropods	Bivalves	Cephalopods
Where do they live?			
How many shells?			
Examples			

Value of Mollusks

I found this information on page _____.

Organize the uses of mollusks *and the* problems they cause *by completing the chart below.*

Uses of Mollusks

_____
_____
_____

Problems Mollusks Cause

_____
_____
_____

CONNECT IT — Discuss several ways you could protect a boat from being damaged by shipworms.

_____
_____

38 *Mollusks, Worms, Arthropods, Echinoderms*

Mollusks, Worms, Arthropods, Echinoderms

Section 2 Segmented Worms

Skim Section 2 of your book. Write three questions that come to mind. Look for answers to your questions as you read the section.

1. _____
2. _____
3. _____

Review Vocabulary **Define** aerate *using your book or a dictionary.*

aerate _____

New Vocabulary Use your book or a dictionary to define the following terms. Then use each term in a sentence to show its scientific meaning.

setae _____
_____
_____
_____

crop _____
_____
_____
_____

gizzard _____
_____
_____
_____

Academic Vocabulary Use a dictionary to define survive as it might be used in science.

survive _____
_____

Mollusks, Worms, Arthropods, Echinoderms

Name _____ Date _____

Section 2 Segmented Worms (continued)

Main Idea	**Details**

Segmented Worm Characteristics

I found this information on page _____.

Identify characteristics of segmented worms *in the chart below.*

Characteristics of Segmented Worms	
Type of symmetry	
Body description	
Where they live	

Earthworm Body Systems

I found this information on page _____.

Sequence *and define the* functions of an earthworm's digestive system *by completing the flow chart.*

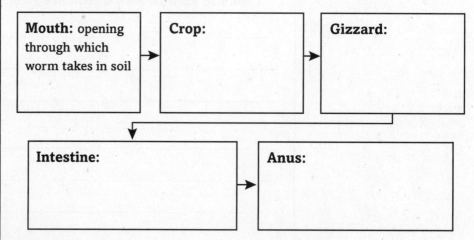

Mouth: opening through which worm takes in soil → **Crop:** → **Gizzard:** → **Intestine:** → **Anus:**

Marine Worms

I found this information on page _____.

Identify three ways that marine worms move.

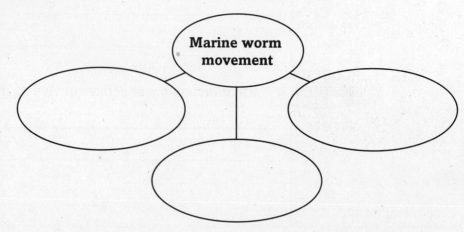

Marine worm movement

140 *Mollusks, Worms, Arthropods, Echinoderms*

Name _____ Date _____

Section 2 Segmented Worms (continued)

Main Idea | Details

Leeches and Leeches and Medicine

I found this information on page _____.

Summarize the process by which leeches feed on the blood of other animals. Then explain how the process is useful in medicine.

_____
_____
_____
_____
_____
_____
_____
_____

Value of Segmented Worms

I found this information on page _____.

Identify ways segmented worms are helpful in the organizer below.

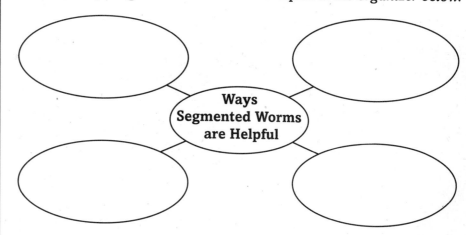

Origin of Segmented Worms

I found this information on page _____.

Compare three similarities of mollusks and worms which suggest that they share a common ancestor.

- _____
- _____
- _____

CONNECT IT Explain why there are not many fossils of ancient worms.

_____
_____

Mollusks, Worms, Arthropods, Echinoderms

Name _____ Date _____

Mollusks, Worms, Arthropods, Echinoderms

Section 3 Arthropods

Scan the What You'll Learn *statements for Section 3 of your book. Identify three topics that will be discussed.*

1. _____
2. _____
3. _____

Review Vocabulary **Define** venom *using your book or a dictionary.*

venom _____

New Vocabulary *Use your book or a dictionary to define the following terms.*

appendage _____
_____

molting _____
_____

spiracle _____
_____

metamorphosis _____
_____

Academic Vocabulary *Use a dictionary to define* individual *as it might be used in science.*

individual _____
_____

Mollusks, Worms, Arthropods, Echinoderms

Name _____ Date _____

Section 3 Arthropods (continued)

Main Idea | Details

Characteristics of Arthropods

I found this information on page _____.

Complete the chart below to identify characteristics of arthropods.

Characteristics of Arthropods	
Type of symmetry	
Body description	
Where they live	

Insects

I found this information on page _____.

Organize information about body regions of insects in the outline.

I. Insect body regions
   A. Parts of the head
      1. _____
      2. _____
      3. _____
   B. Parts of the _____
      1. _____
      2. _____
      3. spiracles
   C. Parts of the _____
      1. _____
      2. _____

Arachnids

I found this information on page _____.

Identify three arachnids and one unique characteristic of each.

Types of Arachnids		

Mollusks, Worms, Arthropods, Echinoderms   143

Name _____ Date _____

Section 3 Arthropods (continued)

Main Idea | Details

Centipedes and Millipedes

I found this information on page _____.

Compare and contrast centipedes *and* millipedes *by completing the Venn diagram below with at least six facts.*

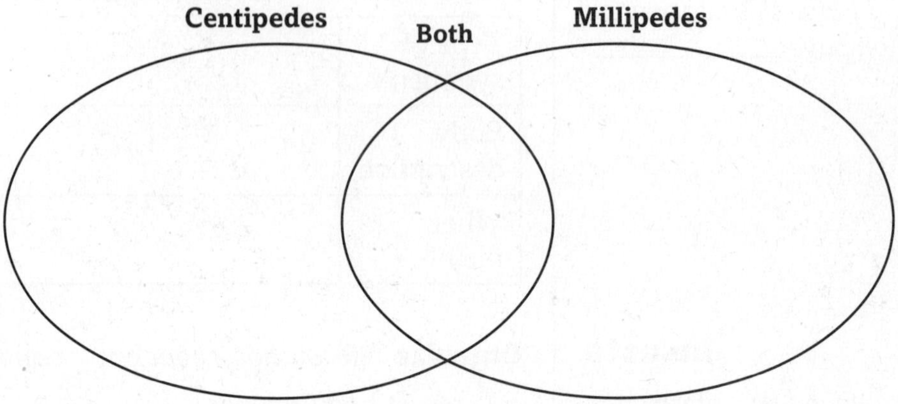

Centipedes Both Millipedes

Crustaceans

I found this information on page _____.

Identify two functions of crustaceans' swimmerets.

1. _____
2. _____

Value of Arthropods

I found this information on page _____.

Summarize helpful functions *and* problems caused by arthropods.

Helpful Arthropod Functions

Problems Arthropods Cause

SYNTHESIZE IT Analyze one method of controlling insect pests. Support your reasoning.

_____
_____
_____

144 *Mollusks, Worms, Arthropods, Echinoderms*

Name _____ Date _____

Mollusks, Worms, Arthropods, Echinoderms

Section 4 Echinoderms

Scan *Section 4 of your book. Use the checklist below.*

☐ Read all the headings.
☐ Read all the bold words.
☐ Look at the charts, graphs, and pictures.
☐ Think about what you already know about echinoderms.

Now, write three things that you want to learn about echinoderms.

1. _____
2. _____
3. _____

Review Vocabulary

Define epidermis *using your book or a dictionary.*

epidermis _____

New Vocabulary

Write a paragraph that explains the meaning and functions of both of the vocabulary terms.

water-vascular system

tube feet

_____
_____
_____
_____
_____
_____

Academic Vocabulary

Use a dictionary to define *network* in a way that it might be used in science.

network _____

Name _____ Date _____

Section 4 Echinoderms (continued)

Main Idea

Echinoderm Characteristics

I found this information on page _____.

Details

Identify characteristics of echinoderms *in the chart below.*

Characteristics of Echinoderms	
Type of symmetry	
Body description	
Where they live	

I found this information on page _____.

Create *a graphic organizer to identify the* functions of a water-vascular system.

146 *Mollusks, Worms, Arthropods, Echinoderms*

Name _____ Date _____

Section 4 Echinoderms (continued)

Main Idea — **Details**

Types of Echinoderms

I found this information on page _____.

Classify the types of echinoderms, *and* **identify** one characteristic of each in the chart below.

Echinoderms	
Type	**Characteristics**
Sea stars	have at least five arms that can regenerate if broken off

Value of Echinoderms

I found this information on page _____.

Summarize *four reasons that* echinoderms are important *to ocean environments.*

1. _____

2. _____

3. _____

4. _____

CONNECT IT Predict in what part of the ocean echinoderms probably live. Support your reasoning.

_____

_____

Mollusks, Worms, Arthropods, Echinoderms 147

Name _____ Date _____

Mollusks, Worms, Arthropods, Echinoderms Chapter Wrap-Up

Review the ideas you listed in the table at the beginning of the chapter. Cross out any incorrect information in the first column. Then complete the table by filling in the third column.

K What I know	W What I want to find out	L What I learned

Review

Use this checklist to help you study.

- ☐ Review the information you included in your Foldable.
- ☐ Study your *Science Notebook* on this chapter.
- ☐ Study the definitions of vocabulary words.
- ☐ Review daily homework assignments.
- ☐ Re-read the chapter and review the charts, graphs, and illustrations.
- ☐ Review the Self Check at the end of each section.
- ☐ Look over the Chapter Review at the end of the chapter.

SUMMARIZE IT After reading this chapter, identify three main ideas that you have learned that you did not know before.

_____
_____
_____
_____

Name _____ Date _____

Fish, Amphibians, and Reptiles

Before You Read

Before you read the chapter, respond to these statements.

1. Write an **A** if you agree with the statement.
2. Write a **D** if you disagree with the statement.

Before You Read	Fish, Amphibians, and Reptiles
	• All vertebrates are chordates.
	• Scales can be used to classify fish.
	• The health of amphibians can indicate the health of the environment.
	• Reptiles must lay their eggs in water.

 Construct the Foldable as directed at the beginning of this chapter.

Science Journal

List two unique characteristics for each animal group you will be studying.

_____
_____
_____
_____
_____
_____
_____

Fish, Amphibians, and Reptiles 149

Name _____ Date _____

Fish, Amphibians, and Reptiles
Section 1 Chordates and Vertebrates

Scan the headings in Section 1 of your book. Predict three topics that will be discussed.

1. _____
2. _____
3. _____

Review Vocabulary **Define** motor responses *using your book or a dictionary.*

motor responses | _____

New Vocabulary *Read the definitions below. Write the correct vocabulary term on the blank to the left of each definition.*

_____ animal that at some point in its development has a notochord, postanal tail, nerve cord, and pharyngeal pouches

_____ pairs of openings between the mouth and the digestive tube found in developing chordates

_____ bones that surround and protect the spinal nerve cord

_____ internal supportive and protective framework found in all vertebrates

_____ tubelike structure that develops into the brain and spinal cord

_____ muscular structure at the end of a developing chordate

_____ flexible, firm structure that extends along the upper part of chordate's body

_____ tough, flexible tissue that joins vertebrae and makes up all or part of the vertebrate endoskeleton

Academic Vocabulary *Use a dictionary to define* external *as it might be used in science.*

external | _____

150 *Fish, Amphibians, and Reptiles*

Name _____ Date _____

Section 1 Chordates and Vertebrates (continued)

Main Idea | Details

Chordate Characteristics

I found this information on page _____.

Model *a developing chordate.* **Label** *its* pharyngeal pouches, postanal tail, notochord, *and* nerve cord.

I found this information on page _____.

Summarize *how the* nerve cord develops *in most chordates.*

_____
_____
_____
_____

Vertebrate Characteristics

I found this information on page _____.

Distinguish *vertebrate chordates from nonvertebrate chordates.* **List** *characteristics of vertebrates that nonvertebrates do not have.*

1. internal framework or endoskeleton
2. _____
3. _____
4. _____
5. _____
6. sometimes have _____

Fish, Amphibians, and Reptiles

Name _____ Date _____

Section 1 Chordates and Vertebrates (continued)

Main Idea | Details

Vertebrate Characteristics

I found this information on page _____.

Identify the 7 main groups of vertebrates.

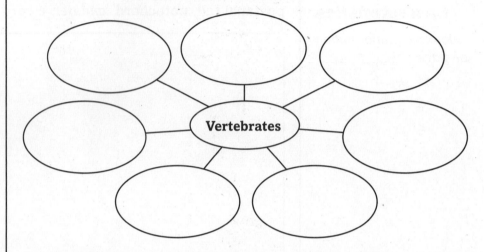

I found this information on page _____.

Define ectotherm *and* endotherm. *Provide a synonym (or word that means the same) and examples for each.*

Ectotherm	Endotherm
Definition:	Definition:
Synonym:	Synonym:
Examples:	Examples:

I found this information on page _____.

Create a timeline to show when vertebrates, amphibians, reptiles, and mammals first appeared. Use a scale of 500 million years ago to the present time.

152 *Fish, Amphibians, and Reptiles*

Name _____ Date _____

Fish, Amphibians, and Reptiles
Section 2 Fish

Skim *Section 2 of your book. Write three questions that come to mind. Look for answers to your questions as you read the section.*

1. _____
2. _____
3. _____

Review Vocabulary **Define** streamline *using your book or a dictionary.*

streamline _____
_____

New Vocabulary *Use your book or a dictionary to define the following terms.*

lateral line _____
_____
_____

fin _____
_____
_____

spawning _____
_____
_____

scales _____
_____
_____

swim bladder _____
_____
_____

Academic Vocabulary *Use a dictionary to define* detect *as it would be used in science.*

detect _____
_____

Fish, Amphibians, and Reptiles

Name _____   Date _____

Section 2 Fish (continued)

Main Idea | Details

Fish Characteristics

I found this information on page _____.

Summarize information about structures and functions of fish fins and scales.

Fins are _____
_____
_____
_____

Scales are _____
_____
_____

I found this information on page _____.

Sequence the steps of fish respiration that take place when a fish obtains oxygen and gets rid of carbon dioxide.

1. A fish takes water into its _____.

2. Water passes over the _____, which contain many

   tiny _____.

3. _____ from the water is exchanged with _____

   _____ from the blood.

4. Water containing _____ passes out through openings on the sides of the fish.

I found this information on page _____.

Compare internal and external fertilization in fish by completing the Venn diagram with at least three facts.

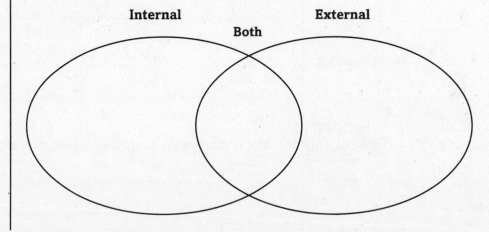

154 *Fish, Amphibians, and Reptiles*

Name _____ Date _____

Section 2 Fish (continued)

Main Idea — **Details**

Types of Fish

I found this information on page _____.

Organize *information about the 3 groups of fish by completing the chart.*

The Three Groups of Fish		
Group	Description	Examples
Jawless fish		
Jawed cartilaginous fish		
Bony fish		

I found this information on page _____.

Model *the body of a typical bony fish by sketching a cutaway view of one. Label its* nostrils, mouth, gills, brain, heart, liver, stomach, intestine, scales, bony vertebrae, *and* swim bladder.

CONNECT IT Analyze how other organisms in a lake might be affected if all the fish living in it disappeared.

_____
_____
_____

Fish, Amphibians, and Reptiles

Name _____ Date _____

Fish, Amphibians, and Reptiles
Section 3 Amphibians

Scan the What You'll Learn statements for Section 3 of your book. Identify three topics that will be discussed.

1. _____
2. _____
3. _____

Review Vocabulary

habitat

Define habitat using your book or a dictionary.

_____
_____
_____

New Vocabulary

Read the definitions below. Write the correct vocabulary term on the blank to the left of each definition.

_____ inactivity in hot, dry months

_____ developmental process in which most amphibians change their body form to become adults

_____ time of inactivity and slowed metabolism during cold weather

_____ species whose overall health reflects the health of the ecosystem in which it lives

Academic Vocabulary

contact

Use a dictionary to define contact as it might be used in science. Then write a sentence that includes the term.

_____
_____
_____
_____

156 Fish, Amphibians, and Reptiles

Name _____ Date _____

Section 3 Amphibians (continued)

Main Idea	Details
Amphibian Characteristics *I found this information on page _____.*	**Complete** *the chart about* amphibians. | Amphibians	
Definition	
Origin of name	
Examples		
I found this information on page _____.	**Compare and contrast** amphibian hibernation *with* estivation *by completing the Venn diagram with at least four facts.* 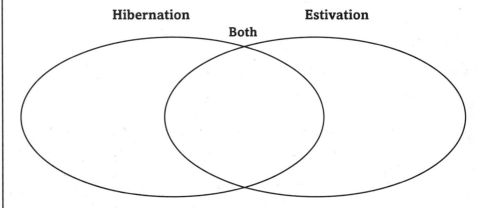

I found this information on page _____.	**Summarize** amphibian respiration and circulation *in the outline.*
	  I. Gas exchange
	     A. Skin is thin, _____, and lined with _____.
	     B. Lungs are small and _____.
	  II. Three-chambered heart
	     A. First chamber _____.
	     B. Second chamber _____.
	     C. Third chamber _____.

Fish, Amphibians, and Reptiles

Name _____ Date _____

Section 3 Amphibians (continued)

Main Idea — Details

Amphibian Characteristics

I found this information on page _____.

Sequence reproduction *and* development in amphibians.

- Eggs are laid in _____ and fertilized _____.
- → Eggs hatch into _____ with fins, _____.
- → Tadpoles develop legs, _____.
- → Adults can live on _____.

Frogs and Toads and Salamanders

I found this information on page _____.

Classify amphibians *by completing the chart.*

Amphibian Groups	Frogs and Toads	Salamanders and Newts
Body structure		
Feeding habits		
Reproduction		

Importance of Amphibians

I found this information on page _____.

Identify *four ways that amphibians are important to humans.*

Why Amphibians are Important

☐ ☐ ☐ ☐

CONNECT IT Think about where amphibians spend their lives. Analyze how this might make them important biological indicators.

_____
_____

158 *Fish, Amphibians, and Reptiles*

Name _____ Date _____

Fish, Amphibians, and Reptiles
Section 4 Reptiles

Skim Section 4 of your book. Write three questions that come to mind. Look for answers to your questions as you read the section.

1. _____

2. _____

3. _____

Review Vocabulary

Define bask *using your book or a dictionary.*

bask _____
_____
_____

New Vocabulary

Use your book or a dictionary to define the vocabulary term. Then use the term in a sentence that shows its scientific meaning.

amniotic egg _____
_____
_____
_____
_____

Academic Vocabulary

Use a dictionary to define interpret *as it might be used in science.*

interpret _____
_____
_____

Name _____ Date _____

Section 4 Reptiles (continued)

Main Idea — Details

Reptile Characteristics

I found this information on page _____.

Summarize reptiles *by completing the chart.*

Characteristics of Reptiles	
Characteristic	**Description or Function**
Skin	
Scales	
Movement	
Body Temperature	
Circulation	
Respiration	

I found this information on page _____.

Model *the structure of the* amniotic egg. *Label the* embryo, shell, yolk sac, egg membrane, *and* air space.

160 *Fish, Amphibians, and Reptiles*

Name _____ Date _____

Section 4 Reptiles (continued)

Main Idea | Details

Types of Modern Reptiles

I found this information on page _____.

Complete *the outline about the* major groups of modern reptiles.

I. Lizards
   A. Body:
      1. Jaw has _____
      2. Toes have _____
   B. Feeding: eat _____

II. Snakes
   A. Jaw:
      1. Has joint that _____
      2. Lower jaw bone used to _____
   B. Have no legs

III. Turtles
   A. Body:
      1. Jaw is _____
      2. Shell consists of _____
   B. Feeding: eat _____

IV. Crocodilians
   A. Body:
      1. Shape is _____.
      2. Head
         a. Crocodile: _____
         b. Alligator: _____
         c. Gavial: _____

The Importance of Reptiles

I found this information on page _____.

   B. Feeding: eat _____

V. The Importance of Reptiles
   A. _____
   B. _____

SUMMARIZE IT

Identify three reptile adaptations that help them survive on land.

_____
_____

Fish, Amphibians, and Reptiles **161**

Name _____   Date _____

Fish, Amphibians, and Reptiles
Chapter Wrap-Up

Now that you have read the chapter, think about what you have learned and complete the table below. Compare your previous answers with these.

1. Write an **A** if you agree with the statement.
2. Write a **D** if you disagree with the statement.

Fish, Amphibians, and Reptiles	After You Read
• All vertebrates are chordates.	
• Scales can be used to classify fish.	
• The health of amphibians can indicate the health of the environment.	
• Reptiles must lay their eggs in water.	

Review
Use this checklist to help you study.

☐ Review the information you included in your Foldable.
☐ Study your *Science Notebook* on this chapter.
☐ Study the definitions of vocabulary words.
☐ Review daily homework assignments.
☐ Re-read the chapter and review the charts, graphs, and illustrations.
☐ Review the Self Check at the end of each section.
☐ Look over the Chapter Review at the end of the chapter.

SUMMARIZE IT After reading this chapter, identify three main ideas that you have learned that you did not know before.

_____
_____
_____
_____

Name _____ Date _____

Birds and Mammals

Before You Read

Before you read the chapter, respond to these statements.

1. Write an **A** if you agree with the statement.
2. Write a **D** if you disagree with the statement.

Before You Read	**Birds and Mammals**
	• A bird has a crop instead of a stomach.
	• Wings are important for nonflying birds.
	• Marsupials are mammals that lay eggs.
	• Bats help pollinate flowers.

Construct the Foldable as directed at the beginning of this chapter.

Science Journal

List similar characteristics of a mammal and a bird. What characteristics are different?

_____
_____
_____
_____
_____
_____
_____

Name _____ Date _____

Birds and Mammals
Section 1 Birds

Scan the headings in Section 1. Identify three topics that will be discussed.

1. _____
2. _____
3. _____

Review Vocabulary **Define** thrust *using your book or a dictionary.*

thrust _____
_____

New Vocabulary *Use your book or a dictionary to define the following terms. Then use each term in a sentence to show its scientific meaning.*

contour feather _____
_____
_____
_____

endotherm _____
_____
_____
_____

preening _____
_____
_____
_____

Academic Vocabulary *Use a dictionary to define* migrate *to reflect its scientific meaning.*

migrate _____
_____

164 *Birds and Mammals*

Name _____ Date _____

Section 1 Birds (continued)

Main Idea

Bird Characteristics

I found this information on page _____.

Details

Complete *the graphic organizer below with three* **common bird** *characteristics.*

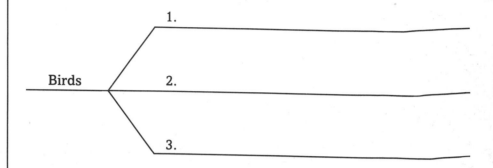

I found this information on page _____.

Summarize *how each structure of a bird's body is* **adapted for** *flight.* ***Complete the chart.***

Adaptations for Flight	
Adaptation	**Description**
Skeleton	
Contour feathers	
Wings	

Birds and Mamm

Name _____ Date _____

Section 1 Birds (continued)

Main Idea	**Details**

Body Systems

I found this information on page _____.

Sequence *the steps in a bird's digestive process in the flow chart.*

| Food is taken into _____ | → | Enters _____ unchewed; there it _____ |

↓

| Moves to _____, where it is _____ | → | Moves to _____, where it is _____ |

↓

| Travels through _____, where nutrients _____ |

I found this information on page _____.

Summarize *how birds' respiratory and circulatory systems provide muscles with sufficient oxygen.*

Respiratory System	Circulatory System

The Importance of Birds

I found this information on page _____.

Summarize *three ways birds positively affect human life.*

1. _____
2. _____
3. _____

SYNTHESIZE IT — List at least three products used in homes that come from birds.

_____
_____

Birds and Mammals

Name _____ Date _____

Birds and Mammals
Section 2 Mammals

Skim *Section 2 of your book. Write three questions that come to mind. Look for answers to your questions as you read the section.*

1. _____
2. _____
3. _____

Review Vocabulary **Define** gland *using your book or a dictionary.*

gland _____

New Vocabulary *Use your book to define the following terms.*

mammary gland _____

gestation period _____

umbilical cord _____

carnivore _____

herbivore _____

omnivore _____

Academic Vocabulary *Use a dictionary to define* attach *to reflect its scientific meaning.*

attach _____

Birds and Mammals 167

Name _____ Date _____

Section 2 Mammals (continued)

Main Idea — **Details**

Characteristics of Mammals

I found this information on page _____.

Create *a graphic organizer to identify at least four characteristics of mammals.*

Body Systems

I found this information on page _____.

Summarize *mammal body systems. Write two facts for each.*

| Mammal Body Systems ||
System	Description
Circulatory	
Respiratory	
Nervous	
Digestive	

Name _____ Date _____

Section 2 Mammals (continued)

Main Idea	**Details**

Types of Mammals

I found this information on page _____.

Compare the 3 types of mammals by completing the chart below.

Types of Mammals		
Type	How Bear Young	Example
Monotremes		
	give birth to immature young that usually crawl into pouch on female's abdomen	
		human

Importance of Mammals

I found this information on page _____.

Complete the outline below.

A. Mammals help keep balance in the ecosystem

1. _____

2. _____

B. Some mammals are in danger

1. _____

2. _____

CONNECT IT A drought kills many of the plants upon which the local herbivores rely upon. Might this affect the local carnivores as well? Explain.

_____

_____

_____

Birds and Mammals 169

Name _____ Date _____

Birds and Mammals Chapter Wrap-Up

Now that you have read the chapter, think about what you have learned and complete the table below. Compare your previous answers with these.

1. Write an **A** if you agree with the statement.
2. Write a **D** if you disagree with the statement.

Birds and Mammals	**After You Read**
• A bird has a crop instead of a stomach.	
• Wings are important for nonflying birds.	
• Marsupials are mammals that lay eggs.	
• Bats help pollinate flowers.	

Review
Use this checklist to help you study.

☐ Review the information you included in your Foldable.
☐ Study your *Science Notebook* on this chapter.
☐ Study the definitions of vocabulary words.
☐ Review daily homework assignments.
☐ Re-read the chapter and review the charts, graphs, and illustrations.
☐ Review the Self Check at the end of each section.
☐ Look over the Chapter Review at the end of the chapter.

SUMMARIZE IT After reading this chapter, identify three key facts that you have learned that you did not know before.

_____
_____
_____
_____

Name _____ Date _____

Animal Behavior

Before You Read

Before you read the chapter, respond to these statements.

1. Write an **A** if you agree with the statement.
2. Write a **D** if you disagree with the statement.

Before You Read	Animal Behavior
	• A bird must learn how to build a nest.
	• A gosling follows the first moving object it sees after hatching.
	• Some animals may show submissive behavior to prevent another animal from attacking.
	• Many animals move to new locations when the seasons change.

Construct the Foldable as directed at the beginning of this chapter.

Science Journal

What behaviors might an animal use to signal that a territory is occupied?

_____
_____
_____
_____
_____
_____

Animal Behavior
Section 1 Types of Behavior

Skim the *What You'll Learn* statements in Section 1. Predict three topics that you expect will be discussed in this section.

1. _____
2. _____
3. _____

Review Vocabulary — **Define** salivate *to show its scientific meaning.*

salivate _____

New Vocabulary — *Read the definitions below. Write the correct vocabulary terms on the blanks in the left column.*

_____ way an organism interacts with other organisms and its environment

_____ behavior that an organism is born with and that does not need to be learned

_____ animal's formation of a social attachment to another organism during a specific period following birth or hatching

_____ modifying behavior so that a response to one stimulus becomes associated with a different stimulus

_____ form of reasoning that allows animals to use past experiences to solve new problems

Academic Vocabulary — *Use a dictionary to define* internal *to show its scientific meaning.*

internal _____

172 *Animal Behavior*

Name _____ Date _____

Section 1 Types of Behavior (continued)

Main Idea | Details

Behavior

I found this information on page _____.

Complete *the flow charts with examples of internal and external stimuli and responses.*

	Stimulus		Response
External		→	
Internal		→	

Innate Behavior

I found this information on page _____.

Identify *two types of innate behavior. Define them and provide at least two examples of each.*

Innate Behaviors		
Type of Behavior	What It Is	Examples

Animal Behavior 173

Name _____ Date _____

Section 1 Types of Behavior (continued)

Main Idea | Details

Learned Behavior

I found this information on page _____.

Analyze *the importance of* learned behavior *for animals.*

Learned behaviors help animals _____ _____. Animals that can learn are _____ _____ than those that cannot. Learned behavior is most commonly found in animals with _____ life spans.

I found this information on page _____.

Summarize *four ways behaviors are learned.*

Behavior Name: Example:	Behavior Description: An animal forms a social attachment within a short time after birth or hatching.
Behavior Name: Example:	Behavior Description:
Behavior Name: Example:	Behavior Description:
Behavior Name: Example:	Behavior Description:

CONNECT IT Moths move toward light. Cockroaches move away from it. What type of behavior is this? Would these animals be able to change this behavior?

_____
_____
_____

174 *Animal Behavior*

Name _____   Date _____

Animal Behavior
Section 2 Behavioral Interactions

Scan *Section 2 by reading the headings and examining the illustrations. Then write three questions that you hope to answer as you read the section. Look for the answers as you read.*

1. _____
2. _____
3. _____

Review Vocabulary **Define** nectar *to show its scientific meaning.*

nectar _____

New Vocabulary *Use your book to define the following terms. Then use each term in a sentence.*

pheromone _____

cyclic behavior _____

migration _____

Academic Vocabulary *Define* dominate *to show its scientific meaning.*

dominate _____

Animal Behavior   175

Name _____ Date _____

Section 2 Behavioral Interactions (continued)

Main Idea	**Details**

Instinctive Behavior Patterns

I found this information on page _____ .

Identify two instinctive ritual animal behaviors.

1. _____

2. _____

Social Behavior

I found this information on page _____ .

Identify three advantages for animals living in groups.

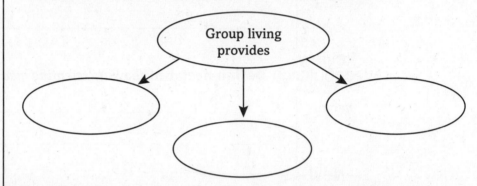

I found this information on page _____ .

Summarize the key features of a *society* in the paragraph below.

A society is _____

_____. Members of

societies have specific roles. In societies that are organized by

dominance, _____

_____.

Territorial Behavior

I found this information on page _____ .

Organize information about territorial behavior. Identify how animals mark their territories and why and how they defend them.

Animal Territories		
Identified by:	Why defended:	How defended:

176 Animal Behavior

Name _____ Date _____

Section 2 Behavioral Interactions (continued)

Main Idea | **Details**

Communication

I found this information on page _____.

Classify *types of* animal communication. **Complete the table below.**

Type of Communication	What It Is	Example
	behaviors that allow males and females of a species to recognize and mate with each other	
Chemical communication		
	Animals make sounds to communicate with other animals of the same species.	
		firefly giving off a flash of light to attract a mate

Cyclic Behavior

I found this information on page _____.

Define *each of the following cyclic behaviors.*

circadian rhythm: _____

_____

hibernation: _____

_____

estivation: _____

_____

Name _____   Date _____

Animal Behavior Chapter Wrap-Up

Now that you have read the chapter, think about what you have learned and complete the table below. Compare your previous answers with these.

1. Write an **A** if you agree with the statement.
2. Write a **D** if you disagree with the statement.

Animal Behavior	After You Read
• A bird must learn how to build a nest.	
• A gosling follows the first moving object it sees after hatching.	
• Some animals may show submissive behavior to prevent another animal from attacking.	
• Many animals move to new locations when the seasons change.	

Review

Use this checklist to help you study.

- ☐ Review the information you included in your Foldable.
- ☐ Study your Science ebook on this chapter.
- ☐ Study the definitions of vocabulary words.
- ☐ Review daily homework assignments.
- ☐ Re-read the chapter and review the charts, graphs, and illustrations.
- ☐ Review the Self Check at the end of each section.
- ☐ Look over the Chapter Review at the end of the chapter.

SUMMARIZE IT After reading this chapter, identify three things that you have learned about animal behavior.

_____
_____
_____
_____
_____
_____

Animal Behavior

Name _____ Date _____

Structure and Movement

Before You Read

Preview the chapter title, section titles, and section headings. Complete the first two columns of the chart by listing at least two ideas for each section in each column.

K What I know	W What I want to find out

 Construct the Foldable as directed at the beginning of this chapter.

Science Journal

Imagine that your body did not have a support system. Describe how you might perform your daily activities.

_____
_____
_____
_____
_____
_____

Structure and Movement
Section 1 The Skeletal System

Skim the headings in Section 1. Write three questions that come to mind about bones and joints.

1. _____
2. _____
3. _____

Review Vocabulary **Define** skeleton *to show its scientific meaning.*

skeleton _____

New Vocabulary *Write the correct vocabulary word next to each definition.*

_____ smooth, slippery, thick layer of tissue that covers the ends of bones

_____ tough band of tissue that holds bones together at joints

_____ tough, tight-fitting membrane that covers a living bone's surface

_____ all of the bones in the body

_____ place where two or more bones come together

Academic Vocabulary *Use a dictionary to define* **transfer** *as a verb.*

transfer _____

180 Structure and Movement

Name _____ Date _____

Section 1 The Skeletal System (continued)

Main Idea — Details

Living Bones

I found this information on page _____.

Organize information about the functions of the skeletal system. Complete the concept web.

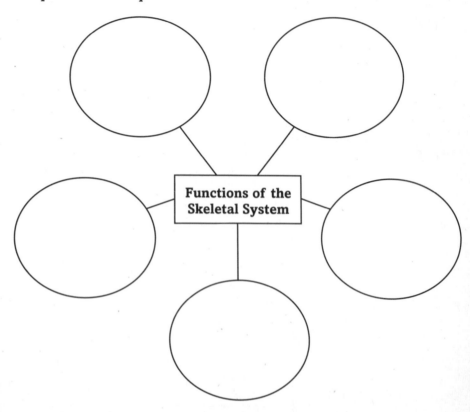

Bone Structure

I found this information on page _____.

Summarize the functions of the following five parts of a bone.

Periosteum: _____
_____

Compact bone: _____
_____

Spongy bone: _____
_____

Marrow cavity: _____
_____

Cartilage: _____
_____

Structure and Movement

Name _____ Date _____

Section 1 The Skeletal System (continued)

Main Idea	**Details**

Bone Formation

I found this information on page _____.

Sequence the steps of bone formation.

1. _____
2. _____
3. _____

Joints

I found this information on page _____.

Classify the five types of joints. Describe and give an example of each.

Type	Description	Example
Immovable		
Pivot		
Ball-and-socket		
Hinge		
Gliding		

I found this information on page _____.

Analyze the role of cartilage in bone movement and what happens if bones cannot move smoothly.

_____
_____
_____

SYNTHESIZE IT Suppose that the joints in your shoulders were hinge joints. Evaluate how this would change your daily life.

_____
_____
_____

Structure and Movement

Name _____ Date _____

Structure and Movement
Section 2 The Muscular System

Predict *three topics that will be covered in Section 2. Read the section headings, and look at the illustrations to help you make your predictions.*

1. _____

2. _____

3. _____

Review Vocabulary **Define** *bone to show its scientific meaning.*

bone   _____

_____

New Vocabulary *Write the correct vocabulary term next to each definition.*

_____ involuntary striated muscle found only in the heart

_____ muscle that can be consciously controlled

_____ muscle that moves bones

_____ muscle that cannot be consciously controlled

_____ thick band of tissue that attaches muscles to bones

_____ organ that can relax, contract, and provide the force to move bones and body parts

_____ involuntary, nonstriated muscle found in intestines, bladder, blood vessels, and other organs

Academic Vocabulary **Define** *flexible as an adjective.*

flexible   _____

_____

Structure and Movement

Name _____ Date _____

Section 2 The Muscular System (continued)

Main Idea	**Details**

Movement of the Human Body

I found this information on page _____.

Summarize *the role of muscles in the body.*

_____
_____

Contrast *voluntary and involuntary muscles. Complete the chart.*

Muscle Type	Consciously controlled	Examples
Voluntary		
Involuntary		

Your Body's Simple Machines— Levers

I found this information on page _____.

Model *the types of levers found in the human body.*

- Draw each type of lever, and label the fulcrum, load, and direction of force.
- Give an example of where the lever is located in the body.

First-class lever

Example: _____

Second-class lever

Example: _____

Third-class lever

Example: _____

184 Structure and Movement

Name _____ Date _____

Section 2 The Muscular System (continued)

Main Idea — Details

Classification of Muscle Tissue

I found this information on page _____.

Compare and contrast the three types of muscle tissue.

Type of Muscle	Voluntary or Involuntary	Where Found in the Body
Skeletal muscle		
Cardiac muscle		
Smooth muscle		

Working Muscles

I found this information on page _____.

Summarize how muscles work in pairs.

_____

_____

_____

I found this information on page _____.

Sequence how muscles are fueled by filling in the missing words.

Blood carries _____ to your muscle cells.

When your muscles contract, _____ from these molecules is converted to _____ and _____. When the supply of _____ in the muscle is _____, the muscle becomes _____. As the muscle _____, blood brings more _____ to your muscle cells.

CONNECT IT Suppose a woman began riding her bike more regularly instead of watching TV. Evaluate what kinds of changes in her leg muscles she might start seeing. Explain why this occurs.

_____

_____

Structure and Movement

Name _____ Date _____

Structure and Movement
Section 3 The Skin

Preview the What You'll Learn statements for Section 3. Predict three topics that you will study in this section.

1. _____
2. _____
3. _____

Review Vocabulary **Define** vitamin to show its scientific meaning.

vitamin
_____
_____
_____

New Vocabulary Define each vocabulary term.

epidermis
_____
_____

melanin
_____
_____

dermis
_____
_____

Academic Vocabulary Use a dictionary to define layer as a noun. Then find a sentence in Section 3 that uses the term.

layer
_____
_____
_____

Structure and Movement

Name _____ Date _____

Section 3 The Skin (continued)

Main Idea

Your Largest Organ and Skin Structures

I found this information on page _____.

Details

Create *a cross-section drawing of the* **skin.** *Label the following structures.*

- blood vessels
- dermis
- epidermis
- fatty layer
- hairs
- hair follicles
- nerve endings
- oil glands
- sweat gland
- sweat pore

Write *captions summarizing key facts about the* **dermis** *and* **epidermis.**

Dermis: _____
_____
_____

Epidermis: _____
_____
_____

I found this information on page _____.

Analyze *the role of* **melanin** *in the body.*
_____
_____
_____
_____

Structure and Movement 187

Name _____ Date _____

Section 3 The Skin (continued)

Main Idea	Details
Skin Functions I found this information on page _____.	**Distinguish** the five primary functions of the skin. 1. _____ 2. _____ 3. _____ 4. _____ 5. _____
Skin Injuries and Repair I found this information on page _____.	**Summarize** how bruises form. _____ _____ _____
I found this information on page _____.	**Sequence** the steps as a cut heals.

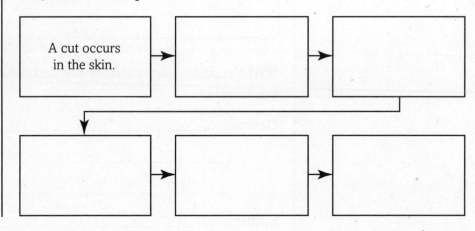

A cut occurs in the skin.

CONNECT IT Analyze why people with severe burns or other damage to large areas of their skin are especially vulnerable to infections. Explain how skin grafts help prevent infections.

_____
_____
_____
_____

Structure and Movement

Tie It Together

Structure and Movement

Design a model that shows how the skeletal and muscular systems work together to allow you to bend your elbow. Present your model to the class and explain how it works.

Name _____ Date _____

Structure and Movement
Chapter Wrap-Up

Review the ideas you listed in the chart at the beginning of the chapter. Cross out any incorrect information in the first column. Then complete the chart by filling in the third column. How do your ideas now compare with those you provided at the beginning of the chapter?

K What I know	W What I want to find out	L What I learned

Review
Use this checklist to help you study.

☐ Review the information you included in your Foldable.
☐ Study your *Science Notebook* on this chapter.
☐ Study the definitions of vocabulary words.
☐ Review daily homework assignments.
☐ Re-read the chapter and review the charts, graphs, and illustrations.
☐ Review the Self Check at the end of each section.
☐ Look over the Chapter Review at the end of the chapter.

SUMMARIZE IT What are the three most important ideas in this chapter?

_____
_____
_____
_____

Name _____ Date _____

Nutrients and Digestion

Before You Read

Before you read the chapter, respond to these statements.

1. Write an **A** if you agree with the statement.
2. Write a **D** if you disagree with the statement.

Before You Read	**Nutrients and Digestion**
	• All foods provide the body with the same amount of energy.
	• What you eat does not affect your health.
	• Sixty percent of your body weight is made up of water.
	• There are bacteria in your digestive tract that make vitamins needed for health.

Construct the Foldable as directed at the beginning of this chapter.

Science Journal

Make a list of all the organs you think are part of your digestive system.

_____
_____
_____
_____
_____
_____
_____

Nutrients and Digestion **191**

Name _____ Date _____

Nutrients and Digestion
Section 1 Nutrition

Skim the headings in Section 1 of this chapter. Write three questions that come to mind.

1. _____
2. _____
3. _____

Review Vocabulary — **Define** molecule *to show its scientific meaning.*

molecule _____
_____

New Vocabulary — *Write a sentence that contains both words in each pair.*

nutrient / food group _____
_____
_____

protein / amino acid _____
_____
_____

carbohydrate / fat _____
_____
_____

vitamin / mineral _____
_____
_____

Academic Vocabulary — *Use a dictionary to define* energy *to show its scientific meaning.*

energy _____
_____

192 *Nutrients and Digestion*

Name _____ Date _____

Section 1 Nutrition (continued)

Main Idea | **Details**

Why do you eat?
I found this information on page _____.

Define calorie *by completing the statement below.*

Calorie: the amount of heat necessary to _____ the temperature of _____ of water _____.

Classes of Nutrients
I found this information on page _____.

Complete *the graphic organizer with key information about* proteins.

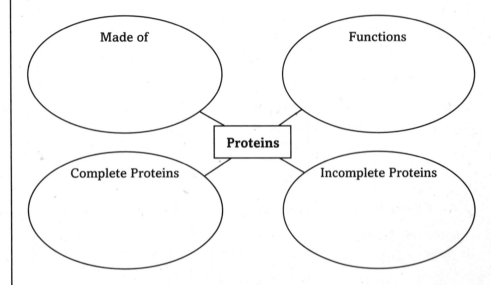

I found this information on page _____.

Compare carbohydrates *and* fats *by completing the chart.*

	Carbohydrates	Fats
Main function(s)		supply energy; help the body absorb vitamins; cushion internal organs
Groups	simple	
Examples		vegetable oils, fats found in meat and animal products

Nutrients and Digestion 193

Name _____ Date _____

Section 1 Nutrition (continued)

Main Idea | Details

I found this information on page _____.

Classify vitamins *by completing the chart.*

Vitamin	Soluble in	Most Beneficial to
A		
B		
C		
D		
E		
K		

I found this information on page _____.

Summarize *why* water *is an important nutrient.*

_____
_____

Food Groups

I found this information on page _____.

Model serving size *for different food categories.*

Group	Servings per Day	Serving Size
bread and cereal		
fruits		
vegetables		
milk		
meat, poultry, fish, beans, eggs		

CONNECT IT

What is the purpose of the food pyramid?

_____
_____
_____

194 Nutrients and Digestion

Name _____ Date _____

Nutrients and Digestion
Section 2 The Digestive System

Preview Section 2 by restating the What You'll Learn statements as questions. Answer each question as you study.

1. _____

2. _____

3. _____

Review Vocabulary **Define** bacteria *to show its scientific meaning.*

bacteria _____

New Vocabulary Read the definitions below. Write the correct vocabulary term on the blank in the left column.

_____	process that breaks down food into small molecules
_____	breakdown of food through chewing, mixing, and churning
_____	occurs when chemical reactions break down large molecules of food into smaller ones
_____	protein that speeds up chemical reactions in the body
_____	muscular contractions that move food through the digestive tract
_____	watery liquid released by the stomach to the small intestine
_____	fingerlike projections covering the wall of the small intestine

Academic Vocabulary *Use a dictionary to define* area *to show its scientific meaning.*

area _____

Nutrients and Digestion 195

Name _____ Date _____

Section 2 The Digestive System (continued)

Main Idea	Details
Functions of the Digestive System *I found this information on page _____.*	**Identify** the four stages of processing food that occur in the human body. 1. _____ 3. _____ 2. _____ 4. _____
Enzymes *I found this information on page _____.*	**Organize** information about digestive enzymes.

Enzyme	Role in digestion
Amylase	
	helps break down proteins
Pancreatic enzymes	

Main Idea	Details
Organs of the Digestive System *I found this information on page _____.*	**Draw** and label the parts of the human digestive system. • Color the organs through which food passes one color. • Color the accessory organs another color. Include the: tongue, mouth, rectum, small intestine, pancreas, anus, stomach, gallbladder, liver, large intestine, esophagus, *and* salivary glands.

Name _____ Date _____

Section 2 The Digestive System (continued)

Main Idea

I found this information on page _____.

Details

Organize information about what happens in the digestive tract.

- List the sections of the digestive tract in the first column.
- Place a checkmark in the appropriate columns showing what occurs in each section.

Section of Digestive Tract	What Occurs		
	Mechanical Digestion	Chemical Digestion	Absorption

Bacteria Are Important

I found this information on page _____.

Complete the table on two types of essential vitamins made by bacteria in the digestive tract.

Vitamin	Function in Body
Vitamin K	
B vitamins	

ANALYZE IT Choose one organ of the digestive system and describe its role in digestion.

_____
_____
_____
_____
_____

Nutrients and Digestion 197

Name _____   Date _____

Nutrients and Digestion   Chapter Wrap-Up

Now that you have read the chapter, think about what you have learned and complete the table below. Compare your previous answers with these.

1. Write an **A** if you agree with the statement.
2. Write a **D** if you disagree with the statement.

Nutrients and Digestion	**After You Read**
• All foods provide the body with the same amount of energy.	
• What you eat does not affect your health.	
• Sixty percent of your body weight is made up of water.	
• There are bacteria in your digestive tract that make vitamins needed for health.	

Review

Use this checklist to help you study.

☐ Review the information you included in your Foldable.
☐ Study your *Science Notebook* on this chapter.
☐ Study the definitions of vocabulary words.
☐ Review daily homework assignments.
☐ Re-read the chapter and review the charts, graphs, and illustrations.
☐ Review the Self Check at the end of each section.
☐ Look over the Chapter Review at the end of the chapter.

SUMMARIZE IT List three important ideas in the chapter.

_____
_____
_____

Name _____ Date _____

Circulation

Before You Read

Before you read the chapter, respond to these statements.

1. Write an **A** if you agree with the statement.
2. Write a **D** if you disagree with the statement.

Before You Read	Circulation
	• The human heart has four chambers.
	• Arteries are blood vessels that carry blood to the heart.
	• Platelets are cell fragments that help fight bacteria and viruses.
	• Lymphatic vessels are like veins in that they have valves.

 Construct the Foldable as directed at the beginning of this chapter.

Science Journal

Infer how the circulatory system provides your body with the nutrients it needs to stay healthy.

_____
_____
_____
_____
_____
_____

Name _____ Date _____

Circulation
Section 1 The Circulatory System

Scan *Section 1 of your book. Read the headings and look at the illustrations. Predict three things that will be discussed.*

1. _____

2. _____

3. _____

Review Vocabulary **Define** *heart using your book or a dictionary.*

heart _____
       _____

New Vocabulary *Read the definitions below. Write the correct vocabulary terms on the blanks in the left column.*

_____ two upper chambers of the heart that contract at the same time

_____ two lower chambers of the heart that contract at the same time

_____ flow of blood to and from the tissues of the heart

_____ flow of blood through the heart to the lungs and back to the heart

_____ flow of blood from the heart to all of the organs and body tissues, except the heart and lungs, with oxygen-poor blood returning to the heart

_____ blood vessel that carries blood away from the heart

_____ blood vessel that carries blood back to the heart

_____ microscopic blood vessel that connects arteries and veins

Academic Vocabulary *Use a dictionary to define* transport *as it would be used in science.*

transport _____
           _____

200 Circulation

Name _____ Date _____

Section 1 The Circulatory System (continued)

Main Idea	**Details**

How Materials Move Through the Body

I found this information on page _____.

Compare and contrast diffusion *and* active transport *by completing the Venn diagram with at least five facts.*

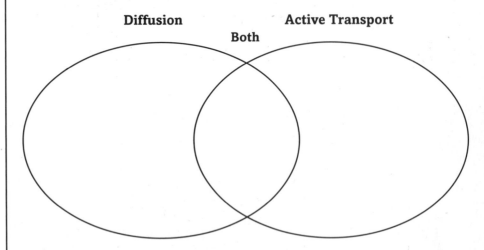

The Heart

I found this information on page _____.

Sequence *the stages in* pulmonary circulation *by completing the flow diagram. Include* aorta, pulmonary veins, pulmonary arteries, right atrium, left atrium, *and* right ventricle.

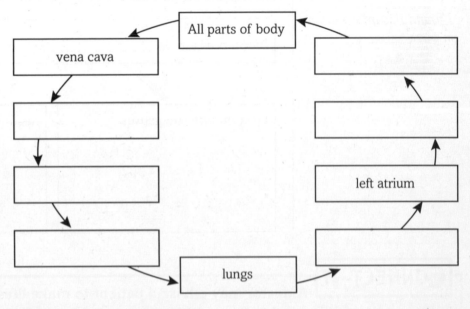

I found this information on page _____.

Summarize *the exchange that occurs between a* systemic capillary *and the* tissue cells *it serves.*

_____
_____
_____

Circulation 201

Name _____ Date _____

Section 1 The Circulatory System (continued)

Main Idea | **Details**

Blood Vessels

I found this information on page _____.

Classify blood vessels by completing the chart.

Blood Vessels		
Type	Function	Description
Arteries		
Capillaries		
Veins		

Blood Pressure

I found this information on page _____.

Define blood pressure and the two numbers used to measure it.

Blood pressure is _____.

First number measures

Second number measures

CONNECT IT A doctor may advise a patient to make lifestyle changes to help prevent cardiovascular disease. Identify several healthful habits the doctor might suggest.

_____
_____
_____

Name _____ Date _____

Circulation
Section 2 Blood

Skim *Section 2 of your book. Write three questions that come to mind. Look for answers to your questions as you read the section.*

1. _____
2. _____
3. _____

Review Vocabulary **Define** blood vessels *using your book or a dictionary.*

blood vessels
_____
_____

New Vocabulary *Use your book or a dictionary to define the following terms.*

platelet
_____
_____
_____

plasma
_____
_____
_____

hemoglobin
_____
_____
_____

Academic Vocabulary *Use a dictionary to define* series *as it would be used in science.*

series
_____
_____
_____

Name _____ Date _____

Section 2 Blood (continued)

Main Idea	Details
Functions of Blood *I found this information on page _____.*	**Create** *a graphic organizer with facts about the* functions of blood.
Parts of Blood *I found this information on page _____.*	**Summarize** *information about the* parts of blood *in the chart below.*

Parts of Blood	
Part	Function
Plasma	
Red blood cells	
White blood cells	
Platelets	

Main Idea	Details
Blood Clotting *I found this information on page _____.*	**Sequence** *the steps in* wound healing *by completing the blanks.* _____ stick to the wound and release _____. Next, _____ forms a sticky net. The net traps _____ and _____ to form a clot. The _____ forms a _____. Then, _____ form under the _____. Finally, the _____ falls off.

204 *Circulation*

Name _____ Date _____

Section 2 Blood (continued)

Main Idea — Details

Blood Types

I found this information on page _____ .

Compare and contrast *the 2 sets of chemical identification tags in blood by completing the Venn diagram with at least five facts.*

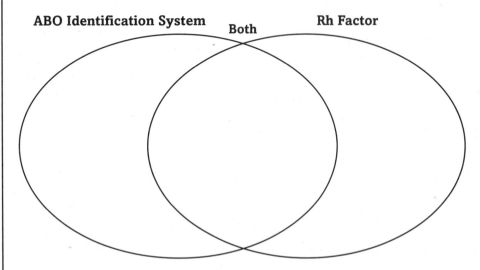

Diseases of Blood

I found this information on page _____ .

Identify *causes and effects of two diseases of the blood.*

	Causes		Effects
Anemia			
	lack of certain vitamins or iron in diet	→	
Leukemia		→	

CONNECT IT Almost immediately after being born, a baby received a blood transfusion of Rh+ blood. Predict the mother's Rh factor. Why did the baby need a blood transfusion?

_____
_____
_____

Circulation 205

Name _____ Date _____

Circulation
Section 3 The Lymphatic System

Scan the What You'll Learn *statements for Section 3 of your book. Identify three topics that will be discussed.*

1. _____
2. _____
3. _____

Review Vocabulary **Define** smooth muscles *using your book or a dictionary.*

smooth muscles _____

New Vocabulary Use your book or a dictionary to define each vocabulary term. Then use the term in a sentence that shows its scientific meaning.

lymph _____
_____
_____

lymphatic system _____
_____
_____

lymphocyte _____
_____
_____

lymph node _____
_____
_____

Academic Vocabulary *Use a dictionary to define* occur *as it would be used in science.*

occur _____

Name _____ Date _____

Section 3 The Lymphatic System (continued)

Main Idea | Details

Functions of the Lymphatic System

I found this information on page _____.

Define tissue fluid *and describe its relationship to the* lymphatic *system.*

| Tissue fluid is _____ _____ _____ _____ _____. | ➔ | The **lymphatic system** collects _____ _____. While in the lymphatic system, the fluid is called _____. |

I found this information on page _____.

Sequence *the stages by which lymph travels through the lymphatic system.*

Tissue fluid enters a network of _____.

⬇

The _____ carry lymph to larger _____.

⬇

The _____ drain into _____ near the heart.

I found this information on page _____.

Summarize *how the* **lymphatic system** *transports lymph. Discuss the role of smooth muscles and valves.*

_____
_____
_____
_____

Circulation 207

Name _____ Date _____

Section 3 The Lymphatic System (continued)

Main Idea	**Details**

Lymphatic Organs

I found this information on page _____.

Model *the* lymphatic system *by drawing it within an outline of the human body. Indicate and label* lymph nodes, lymph vessels, lymphatic duct, thoracic duct, tonsils, thymus, *and* spleen.

A Disease of the Lymphatic System

I found this information on page _____.

Summarize *how* HIV *affects the lymphatic system.*

_____
_____
_____
_____
_____

CONNECT IT Analyze why people who have HIV are at higher risk from the flu or pneumonia than people who are HIV-negative?

_____
_____
_____
_____

208 *Circulation*

Tie It Together

A Checklist for Health

You know that a healthy lifestyle is important for the health of your cardiovascular system.

- Work with a partner to develop a checklist of daily actions to protect your cardiovascular health.
- List actions that are beneficial and actions that should be avoided.
- Provide concrete examples.
- Then make a poster using your checklist.

Name _____ Date _____

Circulation Chapter Wrap-Up

Now that you have read the chapter, think about what you have learned and complete the table below. Compare your previous answers with these.

1. Write an **A** if you agree with the statement.
2. Write a **D** if you disagree with the statement.

Circulation	**After You Read**
• The human heart has four chambers.	
• Arteries are blood vessels that carry blood to the heart.	
• Platelets are cell fragments that help fight bacteria and viruses.	
• Lymphatic vessels are like veins in that they have valves.	

Review

Use this checklist to help you study.

☐ Review the information you included in your Foldable.
☐ Study your *Science Notebook* on this chapter.
☐ Study the definitions of vocabulary words.
☐ Review daily homework assignments.
☐ Re-read the chapter and review the charts, graphs, and illustrations.
☐ Review the Self Check at the end of each section.
☐ Look over the Chapter Review at the end of the chapter.

SUMMARIZE IT After reading this chapter, identify three main concepts that you have learned about circulation.

_____
_____
_____
_____

Name _____ Date _____

Respiration and Excretion

Before You Read

Before you read the chapter, respond to these statements.

1. Write an **A** if you agree with the statement.
2. Write a **D** if you disagree with the statement.

Before You Read	**Respiration and Excretion**
	• Breathing is the process in which the body obtains oxygen and releases energy from food.
	• The respiratory system contains structures that allow humans to speak.
	• If wastes are not removed from the body, they can build up and damage organs.
	• The bladder filters wastes from blood.

 Construct the Foldable as directed at the beginning of this chapter.

Science Journal

How do you think your body adapts to meet your needs while you are playing sports?

_____
_____
_____
_____
_____
_____

Respiration and Excretion **211**

Respiration and Excretion

Section 1 The Respiratory System

Skim the headings of Section 1. Write questions about the respiratory system that you think will be answered in the section.

1. _____
2. _____
3. _____

Review Vocabulary — **Define** lungs to show its scientific meaning.

lungs _____

New Vocabulary — Write four sentences, each containing two of the vocabulary terms. Use each word at least once.

pharynx _____

larynx _____

trachea _____

bronchi _____

alveoli _____

diaphragm _____

emphysema _____

asthma _____

Academic Vocabulary — Use a dictionary to define generate as a verb.

generate _____

212 Respiration and Excretion

Name _____ Date _____

Section 1 The Respiratory System (continued)

‹Main Idea› **‹Details›**

Functions of the Respiratory System

I found this information on page _____.

Classify *each process involved in obtaining, transporting, and using oxygen as* breathing, circulation, *or* respiration.

_____ → Oxygen is supplied to the body.

_____ → Oxygen is transported to body cells.

_____ → Body cells use oxygen and release carbon dioxide.

_____ → Carbon dioxide is transported to lungs.

_____ → Carbon dioxide waste is expelled.

Organs of the Respiratory System

I found this information on page _____.

Summarize *respiratory system structures and functions by completing the chart.*

Structure	Function
	food, liquid, and air share this passage after the nose and mouth
	stops food from entering airway
	directs air through vocal cords
Trachea	
	take air into and out of lungs
Alveoli	

Respiration and Excretion 213

Name _____ Date _____

Section 1 The Respiratory System (continued)

Main Idea

Details

Why do you breathe?

I found this information on page _____.

Model *the processes of* inhaling *and* exhaling *in the boxes below.*

Inhaling	Exhaling

Diseases and Disorders of the Respiratory System

I found this information on page _____.

Summarize *respiratory system diseases and disorders.*

Disease/Disorder	Description
Respiratory infections	
	sometimes caused by bacteria; develops when the bronchial tubes are irritated and swell and too much mucus is produced; lasts for a long time
	disease in which the alveoli enlarge, causing an enzyme that breaks down alveoli walls to be produced; alveoli do not function well and blood receives less oxygen; causes shortness of breath
Lung cancer	
Asthma	

CONNECT IT Identify respiratory diseases and disorders described in this chapter that are related to smoking. List symptoms of these diseases.

_____

_____

_____

214 *Respiration and Excretion*

Name _____ Date _____

Respiration and Excretion
Section 2 The Excretory System

Scan the headings and illustrations in Section 2 to determine three processes that are involved in the urinary system's function.

1. _____
2. _____
3. _____

Review Vocabulary

Define blood to show its scientific meaning.

blood _____

New Vocabulary

Write a paragraph using all seven of the new vocabulary terms. Try to use sentences that show the meaning of each term.

urinary system

urine

kidney

nephron

ureter

bladder

urethra

Academic Vocabulary

Use a dictionary to define remove.

remove _____

Name _____ Date _____

Section 2 The Excretory System (continued)

Main Idea — Details

Functions of the Excretory System

I found this information on page _____.

Complete *the following statement with the words provided.*

damage   illness   wastes   death   toxic

If _____ are not removed from the body, _____ substances build up and _____ organs. Serious _____ or _____ may occur.

The Urinary System

I found this information on page _____.

Model *the urinary system. Draw and label the organs of the urinary system.*

[   ]

Summarize *how blood is processed in the kidneys. Identify substances that pass through the filter and substances that are left behind. Identify the structures involved in each stage.*

First stage: _____

_____

_____

Second stage: _____

_____

_____

216   *Respiration and Excretion*

Name _____ Date _____

Section 2 The Excretory System (continued)

Main Idea	**Details**

I found this information on page _____.

Sequence the structures of the urinary system.

bladder   kidney   ureter   urethra

1. _____ 2. _____ 3. _____ 4. _____

Other Organs of Excretion

I found this information on page _____.

Summarize other processes of excretion.

_____
_____
_____
_____

Urinary Diseases and Disorders

I found this information on page _____.

Analyze the effects of each urinary system problem.

Salt imbalance ⟶ [   ]

Blockage of the ureters and urethra ⟶ [   ]

I found this information on page _____.

Identify information about the diagnoses of urinary diseases.

Disease	Method of Diagnosis
Urinary tract disease	
	change in the urine's color
Diabetes	
	increased amounts of albumin

CONNECT IT Describe how blood helps rid the body of wastes.

_____
_____

Name _____ Date _____

Respiration and Excretion
Chapter Wrap-Up

Now that you have read the chapter, think about what you have learned and complete the table below. Compare your previous answers with these.

1. Write an **A** if you agree with the statement.
2. Write a **D** if you disagree with the statement.

Respiration and Excretion	**After You Read**
• Breathing is the process in which the body obtains oxygen and releases energy from food.	
• The respiratory system contains structures that allow humans to speak.	
• If wastes are not removed from the body, they can build up and damage organs.	
• The bladder filters wastes from blood.	

Review
Use this checklist to help you study.

☐ Review the information you included in your Foldable.
☐ Study your *Science Notebook* on this chapter.
☐ Study the definitions of vocabulary words.
☐ Review daily homework assignments.
☐ Re-read the chapter and review the charts, graphs, and illustrations.
☐ Review the Self Check at the end of each section.
☐ Look over the Chapter Review at the end of the chapter.

SUMMARIZE IT List three processes of excretion described in this chapter.

_____
_____
_____
_____

Control and Coordination

Before You Read

Before you read the chapter, respond to these statements.

1. Write an **A** if you agree with the statement.
2. Write a **D** if you disagree with the statement.

Before You Read	**Control and Coordination**
	• You are subjected to thousands of stimuli every day.
	• The brain is made up of about 10,000 neurons.
	• You can't control reflexes because they occur before you know what has happened.
	• You can smell food because it gives off molecules into the air.

 Construct the Foldable as directed at the beginning of this chapter.

Science Journal

Which senses do you think are involved when you respond to a glass crashing on a tile floor?

_____
_____
_____
_____
_____
_____
_____

Control and Coordination

Section 1 The Nervous System

Scan the headings in Section 1 of your book. Write three questions that come to mind.

1. _____
2. _____
3. _____

Review Vocabulary **Define** response *using its scientific meaning.*

response _____

New Vocabulary Use your book to define the following vocabulary terms.

homeostasis _____
_____

neuron _____
_____

synapse _____
_____

reflex _____
_____

central nervous system _____
_____

peripheral nervous system _____
_____

Academic Vocabulary Use a dictionary to define **coordinate** *using its scientific meaning.*

coordinate _____

Name _____ Date _____

Section 1 The Nervous System (continued)

Main Idea — Details

How the Nervous System Works

I found this information on page _____.

Define stimulus *and describe the relationship between stimuli and the nervous system.*

_____
_____
_____
_____
_____
_____

Nerve Cells

I found this information on page _____.

Sequence *the passage of an* impulse *through a nerve cell. Start with receiving the impulse at a dendrite and end with the part of the nerve cell that carries the impulse to muscles, neurons, and glands.*

dendrite	→		→	

The Central Nervous System

I found this information on page _____.

Organize *information about the parts of the brain and their functions by completing the chart below.*

Part of the brain	Function
Cerebrum	
Cerebellum	
Brain stem	

Describe *the function of the* spinal cord.

Spinal cord: _____
_____

Control and Coordination 221

Name _____ Date _____

Section 1 The Nervous System (continued)

Main Idea | Details

The Peripheral Nervous System

I found this information on page _____.

Compare and contrast the two major parts of the peripheral nervous system by completing the graphic organizer below.

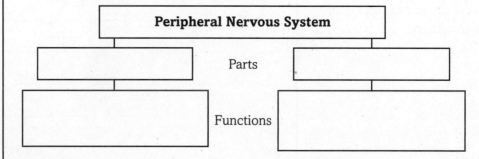

Safety and the Nervous System

I found this information on page _____.

Analyze the diagram of the reflex arc provided in your book. List in order the three neurons involved in the reflex pathway, or arc.

1. _____

2. _____

3. _____

Drugs and the Nervous System

I found this information on page _____.

Distinguish between alcohol and caffeine by completing the Venn diagram with at least two facts for each drug.

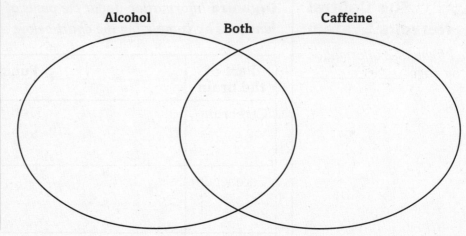

CONNECT IT Infer why alcohol is a controlled substance and caffeine is not.

_____
_____
_____

222 Control and Coordination

Control and Coordination
Section 2 The Senses

Skim the headings of Section 2 to determine the four senses that will be discussed in detail.

1. _____
2. _____
3. _____
4. _____

Review Vocabulary **Define** sense organ *using a dictionary or your book.*

sense organ _____

New Vocabulary Write the correct vocabulary term beside the definition.

_____ light-sensitive tissue at the back of the eye; contains rods and cones

_____ fluid-filled structure in the inner ear in which sound vibrations are converted into nerve impulses that are sent to the brain

_____ nasal nerve cells that become stimulated by molecules in the air and send impulses for interpretation of odors

_____ major sensory receptor on the tongue; contains taste hairs that send impulses for interpretation of tastes

Academic Vocabulary Use a dictionary to define interpret. Use the term in a sentence to show its scientific meaning.

interpret _____

Name _____ Date _____

Section 2 The Senses (continued)

Main Idea	Details
The Body's Alert System *I found this information on page _____ .*	**Create** a graphic organizer to identify three common stimuli that the senses are able to detect.
Vision *I found this information on page _____ .*	**Identify** the functions of each part of the eye.

Part of Eye	Function
Cornea	
Lens	
Retina	
Optic nerve	

Hearing

I found this information on page _____ .

Sequence the parts of the ear in the order that a signal travels.

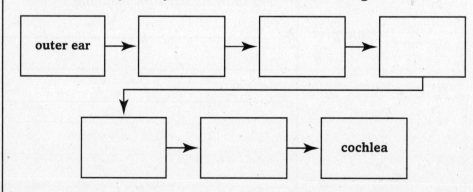

224 Control and Coordination

Name _____ Date _____

Section 2 The Senses (continued)

Main Idea — Details

Smell

I found this information on page _____.

Summarize *how food is smelled by the nose.*

_____
_____
_____
_____
_____

Taste

I found this information on page _____.

Distinguish *the five kinds of tastes in the graphic organizer below.*

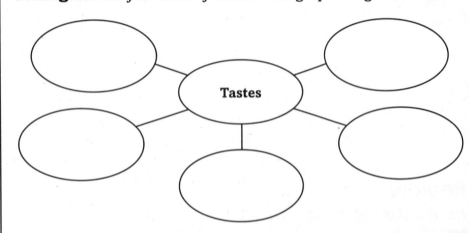

Other Sensory Receptors in the Body

I found this information on page _____.

Summarize *the kinds of stimuli to which the receptors in internal organs and in fingertips can respond by listing them below.*

Internal Organs	Fingertips

EVALUATE IT Identify some advantages of having fingertips and skin with many types of receptors for touch.

_____
_____
_____

Name _____ Date _____

Control and Coordination
Chapter Wrap-Up

Now that you have read the chapter, think about what you have learned and complete the table below. Compare your previous answers with these.

1. Write an **A** if you agree with the statement.
2. Write a **D** if you disagree with the statement.

Control and Coordination	After You Read
• You are subjected to thousands of stimuli every day.	
• The brain is made up of about 10,000 neurons.	
• You can't control reflexes because they occur before you know what has happened.	
• You can smell food because it gives off molecules into the air.	

Review
Use this checklist to help you study.

☐ Review the information you included in your Foldable.
☐ Study your *Science Notebook* on this chapter.
☐ Study the definitions of vocabulary words.
☐ Review daily homework assignments.
☐ Re-read the chapter and review the charts, graphs, and illustrations.
☐ Review the Self Check at the end of each section.
☐ Look over the Chapter Review at the end of the chapter.

SUMMARIZE IT Describe how your nervous system helps protect you.

_____
_____
_____
_____

Name _____ Date _____

Regulation and Reproduction

Before You Read

Before you read the chapter, respond to these statements.

1. Write an **A** if you agree with the statement.
2. Write a **D** if you disagree with the statement.

Before You Read	Regulation and Reproduction
	• Endocrine glands are tissues that produce hormones.
	• Testosterone is the male sex hormone and sperm is the male reproductive cell.
	• Identical twins are not always the same sex.
	• Adulthood is the final stage of human development.

 Construct the Foldable as directed at the beginning of this chapter.

Science Journal

Write a paragraph describing how an emergency call might be handled at a fire station.

_____
_____
_____
_____
_____
_____

Regulation and Reproduction
Section 1 The Endocrine System

Scan the headings, charts, and illustrations in Section 1. Find two glands of the endocrine system that are involved in regulating blood sugar levels and two glands that are involved in regulating calcium levels.

Helps Regulate Blood Sugar Levels	Helps Regulate Calcium Levels

Review Vocabulary

Define tissue *to show its scientific meaning. Then use the word in an original sentence.*

tissue _____

_____

_____

_____

New Vocabulary

Define hormone *to show its scientific meaning.*

hormone _____

_____

_____

_____

Academic Vocabulary

Define distribute *to show its scientific meaning. Then use the word in an original sentence.*

distribute _____

_____

_____

_____

Name _____ Date _____

Section 1 **The Endocrine System** (continued)

Main Idea | Details

Functions of the Endocrine System

I found this information on page _____.

Organize *information about the body's control systems by completing the chart below.*

Body System	Function	Body's Response Time

Endocrine Glands

I found this information on page _____.

Sequence *the events that occur when a gland produces a hormone and sends it to a target tissue.*

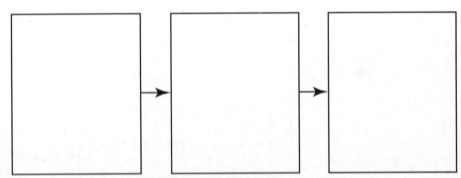

I found this information on page _____.

Distinguish *the four main functions of the endocrine glands by completing the graphic organizer below.*

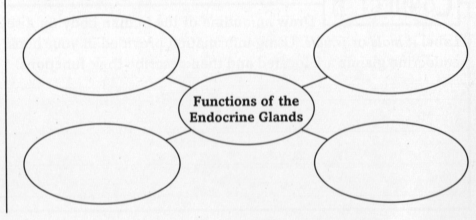

Regulation and Reproduction 229

Name _____   Date _____

Section 1 The Endocrine System (continued)

Main Idea	**Details**

A Negative Feedback System

I found this information on page _____.

Model *a negative-feedback system by completing the cycle chart below.*

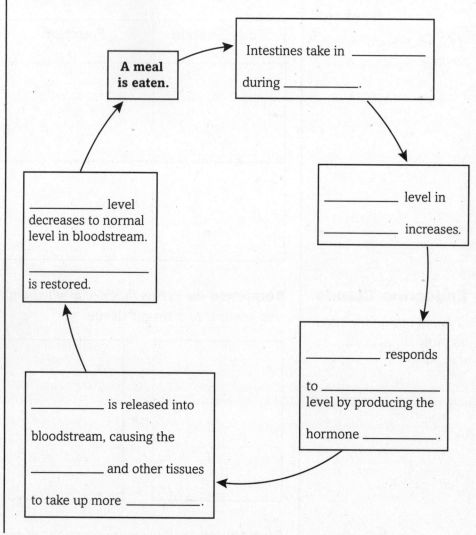

CONNECT IT Draw an outline of the human body on a separate sheet of paper. Label it *male* or *female*. Using information provided in your book, show where endocrine glands are located and then describe their functions.

_____
_____
_____
_____

Regulation and Reproduction
Section 2 The Reproductive System

Predict *three things that might be discussed in Section 2 as you read the headings.*

1. _____

2. _____

3. _____

Review Vocabulary **Define** cilia *as it relates to this section.*

cilia _____
_____

New Vocabulary *Identify the vocabulary terms that match the definitions.*

_____ male organ that produces sperm and testosterone

_____ male reproductive cells

_____ mixture of sperm and a fluid that helps sperm move and supplies the sperm with an energy source

_____ in humans, female reproductive organ that produces eggs

_____ monthly release of an egg from an ovary in a hormone-controlled process

_____ hollow, pear-shaped, muscular organ in which a fertilized egg develops

_____ monthly flow of blood and tissue cells that occurs when the lining of the uterus breaks down and is shed

Academic Vocabulary **Define** respond *using its scientific meaning. Write a sentence that reflects this meaning.*

respond _____
_____
_____

Name _____ Date _____

Section 2 The Reproductive System (continued)

Main Idea

Reproduction and the Endocrine System

I found this information on page _____.

Details

Complete *the graphic organizers below to differentiate the role of the* pituitary gland *in females and males.*

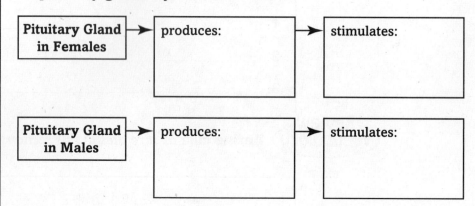

The Male Reproductive System

I found this information on page _____.

Summarize *information about the male reproductive organs in the graphic organizer below.*

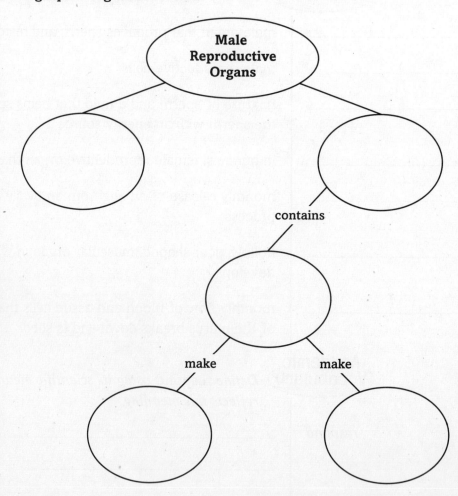

232 Regulation and Reproduction

Name _____ Date _____

Section 2 The Reproductive System (continued)

Main Idea — Details

The Female Reproductive System

I found this information on page _____.

Sequence *the steps through which an egg moves in the* female reproductive system.

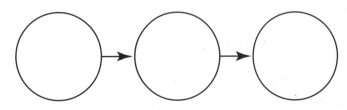

The Menstrual Cycle

I found this information on page _____.

Analyze *the phases of the menstrual cycle, and then complete the chart below.*

	Description	Duration
Phase 1		
Phase 2		
Phase 3 (if fertilized egg does not arrive)		

CONNECT IT Describe how the menstrual cycle would differ in phase 3 if the egg were fertilized. Then infer how future cycles would be affected.

_____
_____
_____
_____
_____

Regulation and Reproduction 233

Regulation and Reproduction
Section 3 Human Life Stages

Skim the headings in Section 3. Then write three questions that you have about human life stages.

1. _____
2. _____
3. _____

Review Vocabulary **Define** nutrient *to show its scientific meaning.*

nutrient _____

New Vocabulary *Define the new vocabulary terms to show their scientific meaning.*

embryo _____

amniotic sac _____

fetus _____

fetus stress _____

Academic Vocabulary **Define** capable. *Use* capable *in an original sentence to show its scientific meaning.*

capable _____

234 Regulation and Reproduction

Name _____ Date _____

Section 3 Human Life Stages (continued)

Main Idea | **Details**

Fertilization
I found this information on page _____.

Sequence *the events that result in the formation of a zygote by completing the following graphic organizer.*

| Sperm enter the vagina and come in contact with chemical secretions in the vagina. |

↓

| |

↓

| |

↓

| |

Multiple Births
I found this information on page _____.

Classify *the following descriptions as applying to either identical twins or fraternal twins. Write either for a description that could fit both categories.*

_____ Two eggs are released and both are fertilized.

_____ A fertilized zygote divides into two separate zygotes.

_____ Twins of the same sex are born.

_____ Twins with different sexes are born.

Development Before Birth
I found this information on page _____.

Create *a time line to indicate when the following events occur: a) embryo forms; b) amniotic sac forms; c) head forms; d) fingers and toes form. Not all weeks will be filled in.*

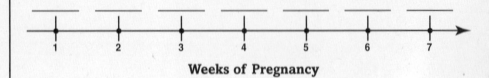

Weeks of Pregnancy

Regulation and Reproduction 235

Name _____ Date _____

Section 3 Human Life Stages (continued)

Main Idea	**Details**

The Birthing Process

I found this information on page _____.

Sequence the events that occur during the birthing process. *The first one has been completed for you.*

1.	Contractions increase.
2.	
3.	
4.	
5.	

Stages After Birth

I found this information on page _____.

Summarize information about the stages after birth *using the chart below.*

Stage	Period in Life	Changes That Occur
Infancy		
Childhood		
Adolescence		
Adulthood		
Older Adulthood		

236 Regulation and Reproduction

Name _____ **Date** _____

Tie It Together

Synthesize It

Create a journal that reflects your own stages of development. Interview your parents to record information about your size at various ages (including birth weight and length) and when you learned certain skills such as the ability to crawl and walk, when you lost your baby teeth, and so on. Try to find pictures of yourself at various ages to include in your journal.

_____
_____
_____
_____
_____
_____
_____
_____
_____
_____
_____
_____
_____
_____
_____
_____
_____
_____
_____
_____
_____
_____
_____

Name _____ Date _____

Regulation and Reproduction
Chapter Wrap-Up

Now that you have read the chapter, think about what you have learned and complete the table below. Compare your previous answers with these.

1. Write an **A** if you agree with the statement.
2. Write a **D** if you disagree with the statement.

Regulation and Reproduction	After You Read
• Endocrine glands are tissues that produce hormones.	
• Testosterone is the male sex hormone and sperm is the male reproductive cell.	
• Identical twins are not always the same sex.	
• Adulthood is the final stage of human development.	

Review

Use this checklist to help you study.

☐ Review the information you included in your Foldable.
☐ Study your *Science Notebook* on this chapter.
☐ Study the definitions of vocabulary words.
☐ Review daily homework assignments.
☐ Re-read the chapter and review the charts, graphs, and illustrations.
☐ Review the Self Check at the end of each section.
☐ Look over the Chapter Review at the end of the chapter.

SUMMARIZE IT Explain how the title "Regulation and Reproduction" fits with the content of this chapter.

_____
_____
_____
_____

Name _____ Date _____

Immunity and Disease

Before You Read

Before you read the chapter, respond to these statements.

1. Write an **A** if you agree with the statement.
2. Write a **D** if you disagree with the statement.

Before You Read	Immunity and Disease
	• Your skin is one of your body's first lines of defense against disease.
	• A vaccine is given to cure a disease.
	• AIDS and HIV are the same thing.
	• You can catch diabetes from another person.

Construct the Foldable as directed at the beginning of this chapter.

Science Journal

Write a paragraph describing a battle between your white cells and a foreign invader.

_____
_____
_____
_____
_____
_____

Immunity and Disease **239**

Immunity and Disease
Section 1 The Immune System

Read the title and headings of the section. Predict two topics that will be discussed in this section.

1. _____

2. _____

Review Vocabulary **Define** enzyme to show its scientific meaning.

enzyme

_____

New Vocabulary Write the vocabulary term that matches each definition.

_____ complex group of defenses that protects the body against pathogens

_____ molecule that is foreign to the body

_____ protein made in response to a specific antigen

_____ immunity in which the body makes its own antibodies in response to an antigen

_____ immunity in which antibodies that have been produced in another animal are introduced to the body

_____ process of giving a vaccine by injection or by mouth

Academic Vocabulary Use a dictionary to define specific to show its scientific meaning.

specific

_____

Name _____ Date _____

Section 1 The Immune System (continued)

Main Idea	**Details**

Lines of Defense

I found this information on page _____.

Summarize *your body's* first-line defense strategies.

| Skin | Respiratory System |

First-line Defenses

| Digestive System | Circulatory System |

I found this information on page _____.

Sequence *what happens when an* antigen *enters the body.*

1. _____

2. _____

3. _____

4. _____

5. _____

Immunity and Disease 241

Name _____ Date _____

Section 1 The Immune System (continued)

Main Idea

I found this information on page _____ .

Details

Contrast active *and* passive immunity. *Complete the chart.*

	Active Immunity	Passive Immunity
What It Is		
How You Get It		
How Long It Lasts		

I found this information on page _____ .

Summarize how a vaccine *helps protect your body against a pathogen. Complete the flow chart.*

A vaccine is injected or given by mouth.

↓

↓

CONNECT IT

Many schools require children to be vaccinated against diseases such as measles before they begin school. Analyze why the schools might have this requirement.

_____
_____
_____

Name _____ Date _____

Immunity and Disease
Section 2 Infectious Diseases

Skim Section 2. Write three questions you would like to have answered. Then look for the answers as you read.

1. _____
2. _____
3. _____

Review Vocabulary **Define** protist *using your book or a dictionary.*

protist _____

New Vocabulary *Use your book to define each vocabulary term.*

pasteurization _____

virus _____

infectious disease _____

biological vector _____

sexually transmitted disease (STD) _____

Academic Vocabulary *Use a dictionary to define* complex *using its scientific meaning.*

complex _____

Name _____ Date _____

Section 2 Infectious Diseases (continued)

Main Idea	Details
Disease in History	**Distinguish** *the important contributions of* Louis Pasteur, Robert Koch, *and* Joseph Lister *to the treatment of infectious diseases.*

I found this information on page _____ .

Pasteur: _____
_____
_____
_____

Koch: _____
_____
_____
_____

Lister: _____
_____
_____
_____

I found this information on page _____ .

Identify *examples of diseases caused by each type of organism.*

Pathogen	Diseases Caused
Bacteria	
Protists	
Fungi	
Viruses	

How Diseases Are Spread

I found this information on page _____ .

Identify *four ways in which diseases can be transmitted.*

1. _____
2. _____
3. _____
4. _____

Immunity and Disease

Name _____ Date _____

Section 2 Infectious Diseases (continued)

Main Idea | **Details**

Sexually Transmitted Diseases

I found this information on page _____.

Identify examples of each type of sexually transmitted disease and list its symptoms and possible effects.

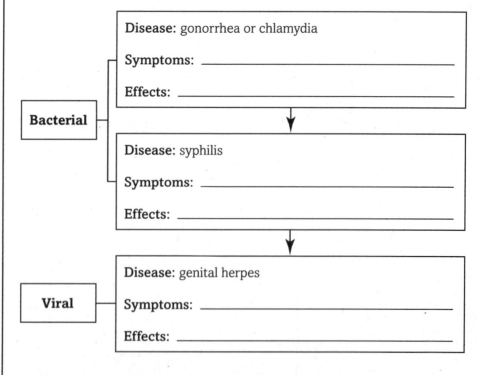

HIV and Your Immune System

I found this information on page _____.

Analyze how HIV harms the immune system. Explain how HIV causes AIDS and what happens when a person has AIDS.

_____
_____
_____
_____
_____
_____

SUMMARIZE IT Describe several things that you can do to prevent infections.

_____
_____
_____

Immunity and Disease

Name _____ Date _____

Immunity and Disease
Section 3 Noninfectious Diseases

Scan the section headings, bold words, and illustrations in Section 3. Write two facts you discovered as you scanned the section.

1. _____
2. _____

Review Vocabulary **Define** gene *using your book or a dictionary.*

gene _____
_____

New Vocabulary *Use your book to define each vocabulary term.*

noninfectious disease _____
_____
_____

allergy _____
_____
_____

allergen _____
_____
_____

chemotherapy _____
_____
_____

Academic Vocabulary *Use a dictionary to define* react. *Then write what you predict* reaction *means. Check your definition in the dictionary.*

react _____
_____
_____

Immunity and Disease

Name _____ Date _____

Section 3 Noninfectious Diseases (continued)

Main Idea — Details

Chronic Disease

I found this information on page _____.

Contrast infectious disease *and* noninfectious disease.

_____
_____
_____
_____

Allergies

I found this information on page _____.

Sequence what happens during an allergic reaction. *Then list some typical symptoms of an allergy.*

| An _____ enters the body. |

↓

| |

↓

| |

Typical symptoms: _____
_____

Diabetes

I found this information on page _____.

Compare and contrast Type 1 *and* Type 2 diabetes. *Complete the chart. Then list common symptoms of both types of diabetes and the possible long-term effects of the disease.*

	Type 1	Type 2
Cause		
Treatment		

Symptoms: _____
_____

Long-term effects: _____
_____

Immunity and Disease

Name _____ Date _____

Section 3 Noninfectious Diseases (continued)

Main Idea | Details

Chemicals and Disease

I found this information on page _____.

Identify the possible harmful effects of the chemicals listed.

Asbestos: _____

Lead-based paints: _____

Alcohol: _____

Cancer

I found this information on page _____.

Summarize information about cancer cells below.

[Diagram: central box labeled "Cancer cells" connected to five empty boxes]

I found this information on page _____.

Summarize the causes, warning signs, and treatments of cancer. Complete the chart.

Causes	
Warning Signs	
Treatments	

CONNECT IT A friend's family has a history of lung and skin cancer. Evaluate some steps your friend could take to reduce his risk of getting these diseases.

_____

_____

Tie It Together

Immunity and Disease

Every winter, many students miss school as a result of colds, influenza, and other infectious diseases. Plan a campaign for your school to teach other students how to reduce their risk of catching these diseases. You might design posters, plan an assembly, or use other ways to get the information out. Outline your plan below.

Name _____   Date _____

Immunity and Disease Chapter Wrap-Up

Now that you have read the chapter, think about what you have learned and complete the table below. Compare your previous answers with these.

1. Write an **A** if you agree with the statement.
2. Write a **D** if you disagree with the statement.

Immunity and Disease	**After You Read**
• Your skin is one of your body's first lines of defense against disease.	
• A vaccine is given to cure a disease.	
• AIDS and HIV are the same thing.	
• You can catch diabetes from another person.	

Review
Use this checklist to help you study.

☐ Review the information you included in your Foldable.
☐ Study your *Science Notebook* on this chapter.
☐ Study the definitions of vocabulary words.
☐ Review daily homework assignments.
☐ Re-read the chapter and review the charts, graphs, and illustrations.
☐ Review the Self Check at the end of each section.
☐ Look over the Chapter Review at the end of the chapter.

SUMMARIZE IT
What are the three most important ideas in this chapter?

_____
_____
_____
_____

Name _____   Date _____

Interactions of Life

Before You Read

Before you read the chapter, respond to these statements.

1. Write an **A** if you agree with the statement.
2. Write a **D** if you disagree with the statement.

Before You Read	Interactions of Life
	• The community includes the top part of Earth's crust, water that covers Earth's surface, and Earth's atmosphere.
	• In nature, most competition occurs between individuals of the same species.
	• Plants and microscopic organisms can move from place to place.
	• Living organisms do not need a constant supply of energy.

 Construct the Foldable as directed at the beginning of this chapter.

Science Journal

Describe how a familiar bird, insect, or other animal depends on other organisms.

_____

_____

_____

_____

_____

_____

Interactions of Life **251**

Name _____ Date _____

Interactions of Life
Section 1 Living Earth

Skim *through Section 1 of your book. Read the headings and look at the figures. Write three questions that come to mind.*

1. _____
2. _____
3. _____

Review Vocabulary **Define** adaptation *using your book or a dictionary.*

adaptation
_____
_____

New Vocabulary *Define each new vocabulary term using your book.*

biosphere
_____
_____

ecology
_____
_____

population
_____
_____

community
_____
_____

habitat
_____
_____

Academic Vocabulary *Define* interact *using a dictionary.*

interact
_____

252 Interactions of Life

Name _____ Date _____

Section 1 **Living Earth** (continued)

Main Idea — Details

The Biosphere

I found this information on page _____.

Complete *this chart to identify three parts of the biosphere.*

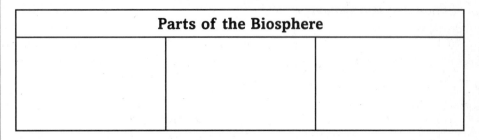

Parts of the Biosphere		

I found this information on page _____.

Contrast *the organisms found in different environments as you complete the concept map. Provide examples of both plants and animals.*

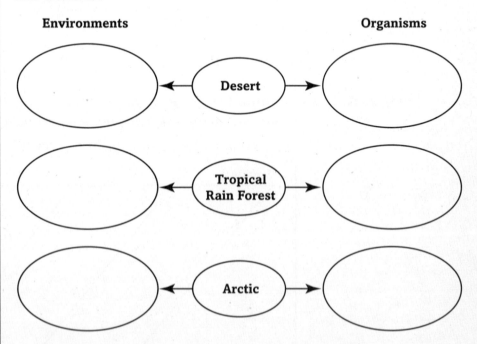

I found this information on page _____.

Analyze *the amount of solar energy that makes Earth the only planet known to support life. Explain why other planets are not suitable for life.*

_____
_____
_____
_____
_____

Interactions of Life 253

Name _____ Date _____

Section 1 Living Earth (continued)

Main Idea | Details

Ecosystems

I found this information on page _____.

Organize the parts of a prairie ecosystem. List three living organisms and three nonliving parts of the ecosystem.

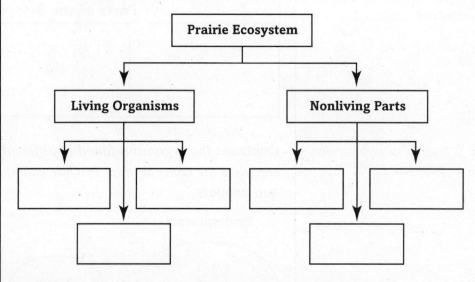

Populations

I found this information on page _____.

Sequence the four levels of organization of living organisms from smallest to largest. Then write an example of each one.

community   organism   ecosystem   population

Smallest   Largest

SYNTHESIZE IT Write about your own life. Use the terms habitat, community, population, and ecosystem to describe your every day interactions.

_____
_____
_____
_____

254 Interactions of Life

Name _____ Date _____

Interactions of Life
Section 2 Populations

Predict *Read the headings in Section 2. Predict three topics that you think will be discussed in this section.*

1. _____

2. _____

3. _____

Review Vocabulary **Define** natural selection *using your book or a dictionary. Then use it in a sentence to show its scientific meaning.*

natural selection _____
_____
_____
_____

New Vocabulary *Create an original sentence using each vocabulary term to show its scientific meaning.*

limiting factor _____
_____

carrying capacity _____
_____

Academic Vocabulary **Define** resource *using a dictionary. Then write a sentence related to the topic of Section 2 using the term.*

resource _____
_____
_____
_____

Interactions of Life 255

Name _____ Date _____

Section 2 Populations (continued)

Main Idea	**Details**

Competition

I found this information on page _____.

Complete the chart below to identify how competing for certain limited resources can affect population growth.

Limited Resource	Why It Limits Population Growth

Population Size

I found this information on page _____.

Compare the two ways of measuring populations by filling in the graphic organizer below.

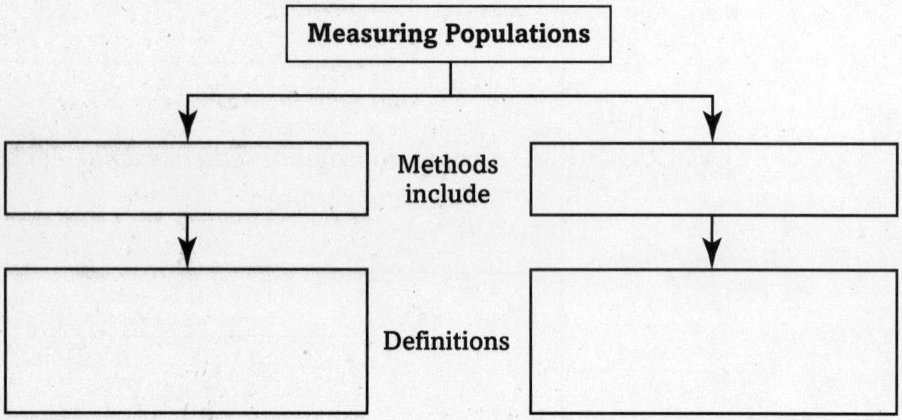

Measuring Populations

Methods include

Definitions

I found this information on page _____.

Contrast carrying capacity *and* biotic potential. *Then identify one factor that can limit each.*

	What It Is	Limiting Factor
Carrying capacity		
Biotic potential		

256 Interactions of Life

Name _____ Date _____

Section 2 Populations (continued)

Main Idea	**Details**

Changes in Populations

I found this information on page _____.

Compare *the effect of differing* birth rates *and* death rates *on population growth as you complete the chart below.*

Population Growth	
Birth Rate Compared to Death Rate	Change in Population
much higher	
slightly higher	
lower	

I found this information on page _____.

Evaluate *the effects of* exponential growth *on a population.*

Size of Population increases → leads to → ◯
◯ ← leads to ←

Summarize *the environmental effects of the exponential growth of a population.*

_____
_____
_____

SYNTHESIZE IT A field is crowded with mice. A new group of mice migrate into the field. Describe how the crowded conditions could affect the mice.

_____
_____
_____
_____
_____
_____

Interactions of Life

Interactions of Life

Section 3 Interactions Within Communities

Scan the What You'll Learn statements for Section 3. Rewrite each statement as a question. As you read the section, try to answer your questions.

1. _____
2. _____
3. _____

Review Vocabulary **Define** social behavior *using your book or a dictionary.*

social behavior _____

New Vocabulary Label each definition with the correct vocabulary term.

_____ an organism that can use an outside energy source like the Sun to make energy-rich molecules

_____ an organism that cannot make its own energy-rich molecules

_____ any close relationship between species

_____ an organism's role in its environment

Academic Vocabulary Define constant *as an adjective. Then use it in a scientific sentence.*

constant _____

Name _____ Date _____

Section 3 Interactions Within Communities (continued)

Main Idea	Details
Obtaining Energy *I found this information on page _____.*	**Compare and contrast** producers *and* consumers *by describing the processes by which each group gets the energy it needs.* 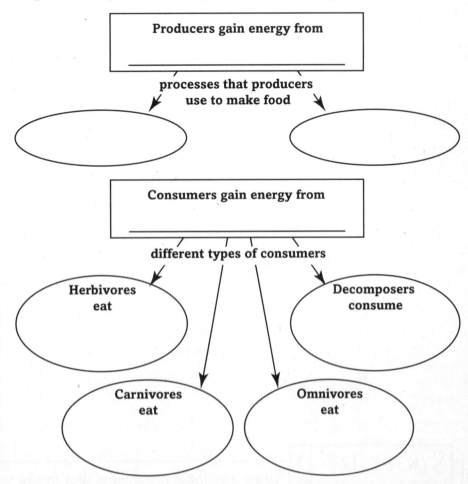
Symbiotic Relationships *I found this information on page _____.*	**Classify** *examples of symbiosis by completing the chart below.*

Type of Symbiosis	Who Benefits?	Example
mutualism		
commensalism		
parasitism		

Interactions of Life 259

Name _____ Date _____

Section 3 Interactions Within Communities (continued)

Main Idea	**Details**

Niches

I found this information on page _____.

Organize *important points about* **niches** *by creating an outline of your reading.*

I. A niche is _____.

　A. how it obtains food

　B. _____

　C. _____

　D. _____

　E. _____

II. Special adaptations that _____
　　can be part of a niche.

　A. **Example:** _____

　　_____

　　_____

　B. **Example:** _____

　　_____

　　_____

SYNTHESIZE IT Draw and label organisms that are in your food chain. Include at least three organisms. Then show how each of these organisms can get the energy it needs.

260 Interactions of Life

Name _____ Date _____

Tie It Together

Observation

Observe the behaviors of a species of animal (for example, squirrels in a park) for at least 15 minutes. Use the chart below to take notes on your observations.

Species:
Date and time of observation:
Number of individuals observed:
Interactions within species:
Food sources observed:
Habitat:
Special adaptations of species:
Interactions observed with other species:

Interactions of Life

Name _____ Date _____

Interactions of Life Chapter Wrap-Up

Now that you have read the chapter, think about what you have learned and complete the table below. Compare your previous answers with these.

1. Write an **A** if you agree with the statement.
2. Write a **D** if you disagree with the statement.

Interactions of Life	**After You Read**
• The community includes the top part of Earth's crust, water that covers Earth's surface, and Earth's atmosphere.	
• In nature, most competition occurs between individuals of the same species.	
• Plants and microscopic organisms can move from place to place.	
• Living organisms do not need a constant supply of energy.	

Review

Use this checklist to help you study.

☐ Review the information you included in your Foldable.
☐ Study your *Science Notebook* on this chapter.
☐ Study the definitions of vocabulary words.
☐ Review daily homework assignments.
☐ Re-read the chapter and review the charts, graphs, and illustrations.
☐ Review the Self Check at the end of each section.
☐ Look over the Chapter Review at the end of the chapter.

SUMMARIZE IT After reading this chapter, identify three things that you have learned about interactions among living organisms.

_____
_____
_____
_____

Name _____ Date _____

The Nonliving Environment

Before You Read

Preview the chapter title, the section titles, and the section headings. List at least two ideas for each section in each column.

K What I know	W What I want to find out

 Construct the Foldable as directed at the beginning of this chapter.

Science Journal

List all the nonliving things that you might see in a picture of a beach, in order of importance. Explain your reasoning for the order you choose.

_____
_____
_____
_____
_____

The Nonliving Environment

Section 1 Abiotic Factors

Preview the What You'll Learn *statements for Section 1. Rewrite each statement into a question.*

1. _____
2. _____
3. _____

Review Vocabulary **Define** environment *to show its scientific meaning.*

environment
_____
_____

New Vocabulary *Define the following terms to show their scientific meanings.*

biotic
_____
_____

abiotic
_____
_____

atmosphere
_____
_____

soil
_____
_____

climate
_____
_____

Academic Vocabulary *Use a dictionary to define* fundamental *as an adjective.*

fundamental
_____
_____

Name _____ Date _____

Section 1 Abiotic Factors (continued)

Main Idea	Details
Environmental Factors *I found this information on page* _____ .	**Classify** *seven environmental factors as biotic or abiotic.*

Factors needed for life	
Biotic	Abiotic
1. _____	1. _____
2. _____	2. _____
	3. _____
	4. _____
	5. _____

Main Idea	Details
Air *I found this information on page* _____ .	**Compare and contrast** *how gases are used during photosynthesis and respiration.*

	Photosynthesis	Respiration
Gas used		
Gas released		
Purpose		

Main Idea	Details
Water and Soil *I found this information on page* _____ .	**Summarize** *how organisms use water and soil. Complete the sentences.* Most organisms are _____ percent water. Processes such as _____, _____, and _____ need water to occur. Environments with plenty of water usually have _____ of organisms than environments with little water. Organisms also need _____. _____, _____, _____, and _____ all live in soil. The type of soil influences the types of _____ that can grow in a region.

The Nonliving Environment 265

Name _____ Date _____

Section 1 Abiotic Factors (continued)

Main Idea — Details

Sunlight
I found this information on page _____.

Label *the diagram to show the flow of energy through living things.* **Label** *consumers, producers,* and *sunlight.*

[_____] → [_____] → [_____]

Temperature
I found this information on page _____.

Analyze *how* latitude *and* elevation *affect temperature.*

Latitude: _____
_____
_____

Elevation: _____
_____
_____

Climate
I found this information on page _____.

Sequence *steps to explain the* rain shadow effect.

1.	Moist air is forced upward by a mountain.
2.	
3.	
4.	

CONNECT IT Describe the climate of your community. Identify its latitude, elevation, temperature, and precipitation characteristics.

_____
_____
_____
_____

Name _____ Date _____

The Nonliving Environment
Section 2 Cycles in Nature

Skim the headings and illustrations in Section 2. List three kinds of cycles you will learn about in the section.

1. _____

2. _____

3. _____

Review Vocabulary **Define** biosphere *to show its scientific meaning.*

biosphere _____

New Vocabulary *Read the definitions below. Write the correct vocabulary term on the blank to the left.*

_____ model describing how carbon molecules move between the living and the nonliving world

_____ process that takes place when a gas changes to a liquid

_____ process in which some types of bacteria in the soil change nitrogen gas into a form of nitrogen that plants can use

_____ process that takes place when a liquid changes to a gas

_____ model describing how water moves from Earth's surface to the atmosphere and back again through evaporation, condensation, and precipitation

_____ model describing how nitrogen moves from the atmosphere to the soil, to living organisms, and then back to the atmosphere

Academic Vocabulary **Define** model *as it is used in the definitions above. Use a dictionary to help you.*

model _____

_____

The Nonliving Environment   267

Name _____ Date _____

Section 2 **Cycles in Nature** (continued)

Main Idea | Details

The Cycles of Matter

I found this information on page _____.

Summarize *the importance of* cycles *to life on Earth.*

_____
_____
_____
_____

The Water Cycle

I found this information on page _____.

Model *the water cycle in a drawing.*

- Label phases of the cycle including evaporation, transpiration, condensation, and precipitation.
- Label the sources and forms the water takes.
- Use arrows to show the direction in which water is moving at each part of the cycle.

The Nitrogen Cycle

I found this information on page _____.

Identify *the three ways that* nitrogen *is made available to plants.*

Plants use nitrogen compounds to build cells.

268 *The Nonliving Environment*

Name _____ Date _____

Section 2 Cycles in Nature (continued)

Main Idea

I found this information on page _____.

Details

Describe *how harvesting removes soil nitrogen and how fertilizer and nitrogen-fixing crops can increase the amount of nitrogen in soil.*

Harvesting: _____
_____

Fertilizer: _____
_____

Nitrogen-fixing crops: _____
_____

The Carbon Cycle

I found this information on page _____.

Model *the carbon cycle. Identify the role of each item shown in the cycle. Draw arrows showing the flow of carbon through the system.*

Air _____
_____.

Producers (Plants and algae)
_____
_____.

Burning wood and fossil fuels
_____
_____.

Consumers _____
_____
_____.

CONNECT IT Choose an organism. Explain its role in the water, nitrogen, and carbon cycles.

_____
_____
_____
_____

Name _____ Date _____

The Nonliving Environment
Section 3 Energy Flow

Skim *Section 3 of your book. Read the headings and look at the illustrations. Write three questions that come to mind.*

1. _____
2. _____
3. _____

Review Vocabulary **Define** energy *to show its scientific meaning.*

energy _____

New Vocabulary *Define the following terms to show their scientific meanings.*

chemosynthesis _____

food web _____

energy pyramid _____

Academic Vocabulary *Use a dictionary to locate the scientific meaning of* convert. *Write a sentence using that scientific meaning.*

convert _____

270   The Nonliving Environment

Name _____ Date _____

Section 3 Energy Flow (continued)

Main Idea	**Details**

Converting Energy

I found this information on page _____.

Compare and contrast photosynthesis *and* chemosynthesis. *Complete the Venn diagram with at least seven points of information from your book.*

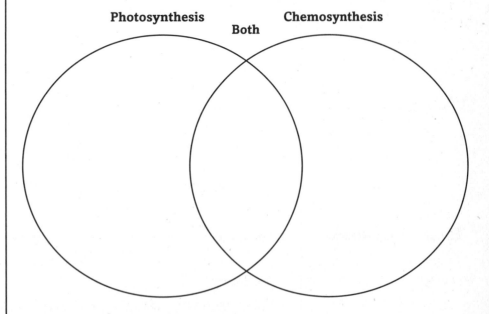

Energy Transfer

I found this information on page _____.

Create *an example of a* food chain.
- Include and label a producer, a herbivore, and a carnivore or omnivore that eats the herbivore.
- Use arrows to show the transfer of energy.

The Nonliving Environment 271

Name _____ Date _____

Section 3 Energy Flow (continued)

Main Idea	Details
I found this information on page _____.	**Synthesize** information about food webs. Draw arrows to show the energy transfers in the food web shown. eagle rattlesnake weasel mouse squirrel plants
Energy Pyramids I found this information on page _____.	**Sequence** the levels of an energy pyramid. • Label each level as containing carnivores, herbivores, or producers. • Label each level with the percentage of total energy that is available at that level. 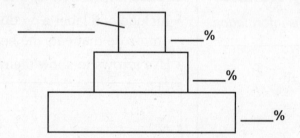

SYNTHESIZE IT Describe the flow of matter and energy in a food chain made up of grasses, mice, and hawks, and what might happen to the food chain if a fire destroyed much of the grass.

_____
_____
_____
_____
_____

272 *The Nonliving Environment*

Name _____ Date _____

Tie It Together

A developer wants to build homes on land near your community and wants to know how the environment will affect the people who live in the homes, and how the homes will affect the environment.

Prepare an environmental study for the developer, including information about

- the abiotic factors in the area that could affect the people in the home
- how the new homes might affect natural cycles and food webs in the area

Use paragraphs and/or pictures to help you explain your points.

The Nonliving Environment

Name _____   Date _____

The Nonliving Environment
Chapter Wrap-Up

Review the ideas you listed in the table at the beginning of the chapter. Cross out any incorrect information in the first column, then complete the table by filling in the third column. How do your ideas compare with those you provided at the beginning of the chapter?

K What I know	W What I want to find out	L What I learned

Review
Use this checklist to help you study.

☐ Review the information you included in your Foldable.
☐ Study your *Science Notebook* on this chapter.
☐ Study the definitions of vocabulary words.
☐ Review daily homework assignments.
☐ Re-read the chapter and review the charts, graphs, and illustrations.
☐ Review the Self Check at the end of each section.
☐ Look over the Chapter Review at the end of the chapter.

SUMMARIZE IT Write three things that you learned while studying this chapter.

_____
_____
_____
_____

Ecosystems

Before You Read

Think about the terms and descriptions below. Infer which term most closely matches the description and write it on the line.

	biome ecosystem estuary intertidal zone
_____	community of living organisms interacting with each other and their physical environment
_____	part of the shoreline that is under water at high tide and exposed to the air at low tide
_____	a large geographic area with an interactive environmental community and similar climate
_____	extremely fertile area where a river meets an ocean; contains a mixture of freshwater and saltwater serves as a nursery for many species

Construct the Foldable as directed at the beginning of this chapter.

Science Journal

What traits might plants on a burning hillside have that enable them to survive and reproduce?

_____
_____
_____
_____
_____
_____

Ecosystems 275

Name _____ Date _____

Ecosystems
Section 1 How Ecosystems Change

Skim *through Section 1 of your text. Write three things that might be discussed in this section.*

1. _____

2. _____

3. _____

Review Vocabulary **Define** *the following key terms using your book or a dictionary.*

ecosystem _____

_____

New Vocabulary

climax community _____

_____

_____

_____

pioneer species _____

_____

_____

_____

succession _____

_____

_____

Academic Vocabulary

stable _____

_____

_____

Name _____ Date _____

Section 1 How Ecosystems Change (continued)

Main Idea	**Details**

Ecological Succession

I found this information on page _____ .

Sequence *the steps in the succession of a lawn to a climax community. The first one has been completed for you.*

	Succession of a Lawn to Climax Community
1.	The grass would get longer.
2.	
3.	
4.	
5.	

I found this information on page _____ .

Organize *the information from your book to compare primary succession with secondary succession.*

	Primary Succession	Secondary Succession
	Lava from a volcano	Fire consumes a forest
Land consists of		
Starts with	_____ break down rock and decay, adding _____.	Soil contains _____.
Animals and wind carry		
Plants add		
Wildlife		

Ecosystems 277

Name _____ Date _____

Section 1 How Ecosystems Change (continued)

Main Idea	**Details**

I found this information on page _____.

Complete the graphic organizer to better understand the characteristics of a climax community.

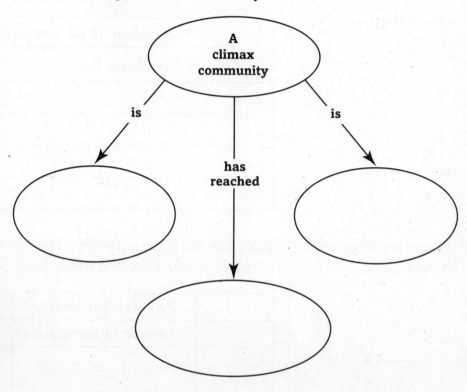

I found this information on page _____.

Identify the three main characteristics of a forest climax community.

1. _____

2. _____

3. _____

CONNECT IT Use the information you have learned about succession to predict the growth of a community in a flooded river basin. Hypothesize whether the succession would be primary succession or secondary succession. Support your answer with facts from your book.

_____
_____
_____
_____

Name _____   Date _____

Ecosystems
Section 2 Biomes

I found this information on page _____.

Analyze Look at the world map of the seven major land biomes in your book. Infer two factors you think scientists might use to classify biomes of the world.

1. _____

2. _____

Review Vocabulary Use the word **climate** *in a scientific sentence.*

climate   _____
           _____

New Vocabulary **Define** Read the definitions below. Write the key terms on the blanks in the left column.

_____   most biologically diverse biome

_____   ideal biome for growing crops and raising cattle and sheep

_____   biome usually having four distinct seasons

_____   cold, dry, treeless biome with a short growing season and permafrost

_____   biome with thin soil where organisms are adapted to survive extreme conditions

_____   biome containing cone-bearing evergreen trees and dense forests

Academic Vocabulary Use a dictionary to define **mature** *as a verb.*

mature   _____
          _____

Ecosystems 279

Name _____ Date _____

Section 2 Biomes (continued)

Main Idea — **Details**

Major Biomes
I found this information on page _____.

Complete the comparison chart using the world map of seven biomes.

	Physical Description	Average Precipitation	Temperature	Location	Plant and Animal Life
Tundra		less than 25 cm per year			Plants: Animals:
Taiga			temperature range: −54°C to 21°C		Plants: Animals:
Temperate Deciduous Forest				eastern US, Europe, parts of Asia and Africa	Plants: Animals:
Temperate Rain Forest	dense forest with a variety of plants and animals				Plants: Animals:

280 *Ecosystems*

Name _____ Date _____

Section 2 Biomes (continued)

Main Idea —————————— **Details** ——————————

	Physical Description	Average Precipitation	Temperature	Location	Plant and Animal Life
Tropical Rain Forest					4 zones of plant and animal life Plants: Animals:
Desert				western US and S. America, Africa, parts of Australia and Asia	Plants: Animals:
Grasslands			mild to hot	prairies—N. America, steppes—Asia, savannas—Africa pampas—S. America	Plants: Animals:

CONNECT IT Analyze the information you recorded about biomes. Compare and contrast the tundra with the desert.

_____
_____
_____
_____

Ecosystems 281

Ecosystems
Section 3 Aquatic Ecosystems

Read the *What You'll Learn* objectives of Section 3. Write questions that come to mind from reading these statements.

1. _____

2. _____

3. _____

Review Vocabulary Define the key terms using your book or a dictionary.

aquatic _____

New Vocabulary

coral reef _____

wetland _____

Academic Vocabulary

promote _____

Freshwater Ecosystems

I found this information on page _____.

Organize the four important factors that determine how well a species can survive in an aquatic environment.

1.
2.
3.
4.

282 *Ecosystems*

Name _____ Date _____

Section 3 Aquatic Ecosystems (continued)

Main Idea | Details

Freshwater Ecosystems

I found this information on page _____.

Compare *fast-moving streams with slower-moving streams as you complete the sentences below about* **freshwater environments.**

Fast-moving Streams

Currents quickly _____
_____.

As water tumbles, air _____.

These streams have clearer _____ and higher

_____.

Slow-moving Streams

Water moves slowly and debris _____.

These environments have higher _____, more

plant _____, and organisms _____

_____

I found this information on page _____.

Classify *each statement as a characteristic of* pond ecosystems, lake ecosystems, *or* both. *Mark* **P** *for pond,* **L** *for lake, or* **B** *for both ecosystems.*

_____ more plants than flowing water environments

_____ deeper water and colder water temperatures

_____ larger body of water

_____ plankton floating near the surface

_____ ecosystem high in nutrients

_____ small, shallow body of water

_____ lower light levels at depth limit types of organisms

_____ plant growth limited to shallow water near shore

_____ water hardly moves

Ecosystems 283

Name _____ Date _____

Section 3 Aquatic Ecosystems (continued)

Main Idea	Details
Freshwater Ecosystems *I found this information on page* _____.	**Organize** information about wetlands *in the concept map.* 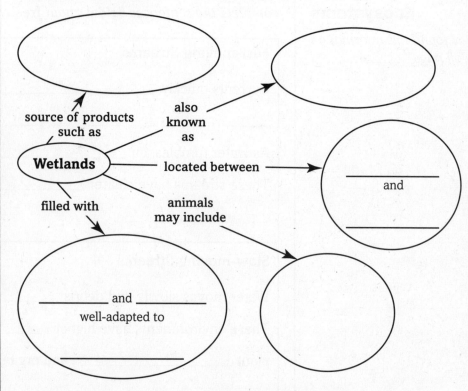

Saltwater Ecosystems
 I found this information on page _____.
	Complete the outline about saltwater ecosystems.
	 I. Coral Reef ecosystems are _____
	 _____
	  A. reefs formed by _____
	  B. damaged by _____
	 II. Seashores
	  A. affected by _____ and _____
	  B. intertidal zone organisms must adapt to _____,
	   _____, and _____ changes
	 III. Estuaries
	  A. contain _____
	  B. are important for _____
	   _____

284 *Ecosystems*

Name _____ Date _____

Tie It Together

Interactions within Ecosystems

Select one of the ecosystems discussed in this chapter. You might choose a tundra ecosystem, a rain forest ecosystem, a coral reef ecosystem, or one of the other ecosystems. Take notes about your ecosystem on the lines below. Then, draw a picture of your ecosystem with its animal and plant inhabitants. Show any interactions that you described in your picture.

My ecosystem is a/an _____.

It includes these plants:
_____
_____

It includes these animals:
_____
_____

Its environment includes these conditions:
_____
_____
_____

Interactions between organisms include these:
_____
_____
_____

Interactions between organisms and the environment include these:
_____
_____
_____
_____

Sketch of My Ecosystem

Ecosystems 285

Ecosystems Chapter Wrap-Up

Think about the terms and descriptions below. Write the term that most closely matches the description on the line in front of the description. Compare your previous responses with these.

	biome ecosystem estuary intertidal zone
_____	community of living organisms interacting with each other and their physical environment
_____	part of the shoreline that is under water at high tide and exposed to the air at low tide
_____	a large geographic area with an interactive environmental community and similar climate
_____	extremely fertile area where a river meets an ocean; contains a mixture of freshwater and saltwater and serves as a nursery for many species

Review

Use this checklist to help you study.

- ☐ Review the information you included in your Foldable.
- ☐ Study your *Science Notebook* on this chapter.
- ☐ Study the definitions of vocabulary words.
- ☐ Review daily homework assignments.
- ☐ Re-read the chapter and review the charts, graphs, and illustrations.
- ☐ Review the Self Check at the end of each section.
- ☐ Look over the Chapter Review at the end of the chapter.

SUMMARIZE IT

After reading this chapter, identify three things that you have learned about ecosystems.

_____
_____
_____
_____
_____

Name _____ Date _____

Conserving Resources

Before You Read

Before you read the chapter, respond to these statements.

1. Write an **A** if you agree with the statement.
2. Write a **D** if you disagree with the statement.

Before You Read	Conserving Resources
	• There is an unlimited supply of fossil fuels.
	• Sun, wind, and heat within Earth's crust can be used to generate power.
	• Acid precipitation washes nutrients from the soil.
	• The ozone layer emits radiation that can harm living cells.

 Construct the Foldable as directed at the beginning of this chapter.

Science Journal

List some resources, other than water, air, and fossil fuels, that we depend on and describe how we use them.

_____
_____
_____
_____
_____
_____

Name _____ Date _____

Conserving Resources
Section 1 Resources

Predict the topics that will be discussed in Section 1 after reading the headings and looking at the illustrations.

1. _____
2. _____
3. _____

Review Vocabulary **Define** geyser to show its scientific meaning.

geyser _____
_____

New Vocabulary Define the following terms to show their scientific meanings.

natural resource _____
_____

hydroelectric power _____
_____

nuclear energy _____
_____

geothermal energy _____
_____

Academic Vocabulary Define modify. Then use it in an original sentence to show its scientific meaning.

modify _____
_____

Name _____   Date _____

Section 1 **Resources** (continued)

Main Idea — Details

Natural Resources

I found this information on page _____.

Compare renewable *and* nonrenewable resources *by completing the chart below.*

Type of Resource	Description	Examples
Renewable		
Nonrenewable		

Fossil Fuels

I found this information on page _____.

Organize information about fossil fuels *in the concept web below.*

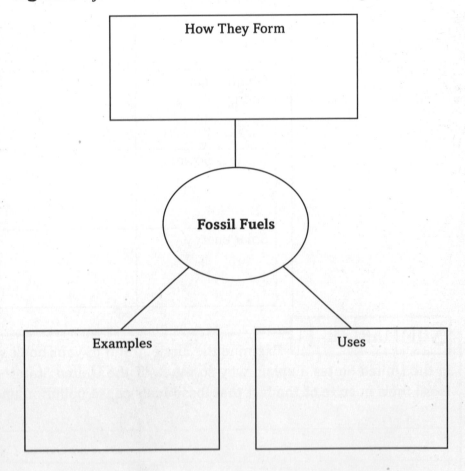

Conserving Resources 289

Name _____ Date _____

Section 1 Resources (continued)

Main Idea

I found this information on page _____.

Details

Summarize three reasons that fossil fuels need to be conserved.

1. _____

2. _____

3. _____

Alternatives to Fossil Fuels

I found this information on page _____.

Organize information about alternative energy resources *below*.

Alternative Energy Resource	Important Information
Hydroelectric power	
Wind energy	
Geothermal energy	
Nuclear power	
Solar energy	

SUMMARIZE IT Examine the circle graph in your book showing energy usage in the United States. Explain why so much of the United States' energy comes from fossil fuels in spite of the fact that fossil fuels cause pollution and are limited in supply.

_____

_____

_____

290   Conserving Resources

Name _____ Date _____

Conserving Resources
Section 2 Pollution

Skim the headings of Section 2 to determine three main types of pollution that will be discussed.

1. _____
2. _____
3. _____

Review Vocabulary **Define** atmosphere *to show its scientific meaning.*

atmosphere _____
_____

New Vocabulary *Read each definition below. Write the correct vocabulary term in the blank to the left.*

_____ substance that contaminates the environment

_____ precipitation that has a pH below 5.6

_____ trapping of heat from the Sun by Earth's atmosphere

_____ waste materials that are harmful to human health or poisonous to living organisms

Academic Vocabulary **Define** affect *to show its scientific meaning.*

affect _____
_____

Name _____ Date _____

Section 2 Pollution (continued)

Main Idea | Details

Acid Precipitation

I found this information on page _____.

Complete the graphic organizer below to identify the effects of *acid rain* and ways to prevent acid rain.

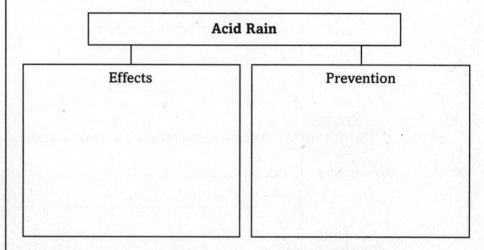

Greenhouse Effect and Ozone Depletion

I found this information on page _____.

Sequence the events that cause the *greenhouse effect* and *ozone depletion* by completing the following graphic organizers.

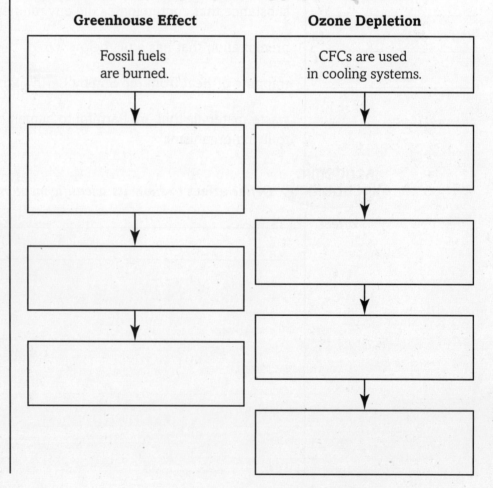

292 *Conserving Resources*

Name _____ Date _____

Section 2 Pollution (continued)

Main Idea | **Details**

Indoor Air Pollution
I found this information on page _____.

Compare and contrast *carbon monoxide and radon as sources of indoor air pollution by completing the following chart.*

Gas	Source	Effect
Carbon monoxide		
Radon		

Water Pollution
I found this information on page _____.

Identify *causes of the following three examples of water pollution.*

1. Surface water pollution: _____
   _____

2. Ocean water pollution: _____
   _____

3. Groundwater pollution: _____
   _____

Soil Loss and Soil Pollution
I found this information on page _____.

Analyze *causes of soil loss and soil pollution.*

A. Causes of soil loss

   1. _____

   2. _____

B. Causes of soil pollution

   1. _____

   2. _____

CONNECT IT Explain in one sentence why people are concerned about pollution.

_____
_____
_____

Conserving Resources

Name _____ Date _____

Conserving Resources
Section 3 The Three Rs of Conservation

Scan the headings of Section 3. List the three Rs of conservation below.

1. _____
2. _____
3. _____

Review Vocabulary

Define the following terms. Then write a paragraph that includes the scientific meaning of all three terms.

reprocessing _____
_____

New Vocabulary

recycling _____
_____

Academic Vocabulary

participate _____
_____

Paragraph: _____
_____
_____
_____
_____
_____
_____

Name _____ Date _____

Section 3 The Three Rs of Conservation (continued)

Main Idea / Details

Conservation

I found this information on page _____.

Identify *reasons for conserving resources by completing the graphic organizer below.*

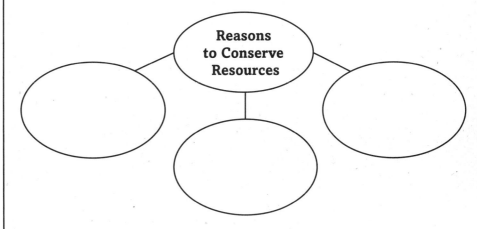

Reduce

I found this information on page _____.

Summarize *four ways to reduce your own use of natural resources.*

1. _____
2. _____
3. _____
4. _____

Reuse

I found this information on page _____.

Define *reusing an item. Then identify at least two examples of ways to reuse items.*

Definition: _____

_____

Examples: _____

_____

_____

_____

_____

Conserving Resources   295

Name _____   Date _____

Section 3 The Three Rs of Conservation (continued)

Main Idea	**Details**

Recycle

I found this information on page _____ .

Summarize recycling *in the following chart.*

Recycling	
Definition:	
Items that can be recycled	
Advantages of recycling	
How recycling is done	

Analyze *the graph that describes the recycling rates of key household items. Then complete the statements.*

The percentages of _____, _____, and _____ being recycled increased from 1990 to 2000.

The percentages of _____, _____, and _____ being recycled decreased from 1995 to 2000.

SYNTHESIZE IT In a small group, discuss why some people do not recycle. Summarize your discussion in the space below.

_____
_____
_____
_____

Conserving Resources

Name _____ **Date** _____

Tie It Together

Conservation

Brainstorm ways to increase the level of conservation practiced in your school. Set a conservation, reuse, or recycling goal. Write a plan to change the school's behavior to meet your goal. If new resources would be needed to implement your plan, hypothesize how you could raise money for what you need.

- Decide which method of conservation you are most concerned about.
- Describe the benefits of practicing that method of conservation in your school.
- Identify practical ways that students can practice conservation.

_____
_____
_____
_____
_____
_____
_____
_____
_____
_____
_____
_____
_____
_____
_____
_____

Name _____ Date _____

Conserving Resources Chapter Wrap-Up

Now that you have read the chapter, think about what you have learned and complete the table below. Compare your previous answers with these.

1. Write an **A** if you agree with the statement.
2. Write a **D** if you disagree with the statement.

Conserving Resources	After You Read
• There is an unlimited supply of fossil fuels.	
• Sun, wind, and heat within Earth's crust can be used to generate power.	
• Acid precipitation washes nutrients from the soil.	
• The ozone layer emits radiation that can harm living cells.	

Review

Use this checklist to help you study.

☐ Review the information you included in your Foldable.
☐ Study your *Science Notebook* on this chapter.
☐ Study the definitions of vocabulary words.
☐ Review daily homework assignments.
☐ Re-read the chapter and review the charts, graphs, and illustrations.
☐ Review the Self Check at the end of each section.
☐ Look over the Chapter Review at the end of the chapter.

SUMMARIZE IT
After reading this chapter, identify three new ways you could practice conservation.
_____
_____
_____
_____

Academic Vocabulary

adapt: to change to fit new conditions

affect: to make something happen; to have an effect on

annual: plant that completes its life cycle in one year

apparent: readily seen, visible, readily understood or perceived; evident; obvious

area: amount or extent of a surface

attach: to be connected

benefit: to help

capable: able to do things; fit

chemical: made by chemistry

chemical bond: the force holding atoms together in a molecule

code: (noun) set of signals representing letters or numerals, used to send messages; (verb) to put in the form or symbols of a code

complex: composed of two or more parts; complicated

compound: (adjective) made of two or more separate parts or elements

constant: not changing; staying the same

contact: act or state of touching or meeting

convert: to change from one form or function to another

coordinate: to cause to work well together

cycle: a complete set of events or phenomena recurring in the same sequence

decline: to weaken or lessen

definite: having exact limits in size, shape, or number of parts

detect: to catch or discover; to manage to perceive

distribute: to divide among several or many

dominate: to control or rule

energy: capacity to perform some type of work or activity

environment: living and nonliving factors that surround an organism

estimate: (noun) an opinion of the value, quality, size, or cost of something; (verb) to form an opinion by reasoning

external: on, or for use on, the outside of the body

facilitate: to make easy or easier

flexible: able to bend or flex

function: (noun) a specific job or purpose; (verb) to carry out a specific action

fundamental: serving as an original or generating source; primary

generate: to originate or bring into existence

hypothesis: something that is suggested as being true for the purposes of argument or of further investigation

identical: same

individual: separate

insert: to put or fit (something) into something else

Life Science

Academic Vocabulary

interact: to act on one another

intermediate: in the middle or being between

internal: of or on the inside

interpret: to tell the meaning of; to understand

involve: to include; to have as part of itself

layer: one thickness of something

mature: to become fully developed or ripe

method: way of doing something; a process

migrate: to move from one place to another place

model: a description used to help visualize something that cannot be directly observed

modify: to undergo change

network: a group of related parts

obtain: to get possession of, especially by some effort

occur: to take place; to be found

participate: to take part; share

physical: having to do with the body

process: series of steps performed in doing something

promote: to contribute to the growth of; to help bring into being

react: to act because something has happened; respond

reject: to refuse to accept or use

relax: to become inactive and lengthen

release: to set free; to let go

remove: to get rid of

require: to be in need of

resource: something used for help or support

respond: to react in response

series: a number of similar things coming one after another

similar: almost, but not exactly the same

soil: mixture of weathered rock, organic matter, water, and air that supports the growth of plant life

source: any person, place, or thing by which something is supplied

specific: exact; particular

stable: firmly established; not changing or fluctuating

structure: arrangement of parts or the way parts are arranged

survive: to continue living

transfer: to convey or transport from one place to another

transport: to carry from one place to another; the act, process, or means of transporting

visible: able to be seen

widespread: widely scattered or prevalent